Springer Undergraduate Mathematics Series

Advisory Board

Other books in this series

Keith E. Hirst

Calculus of One Variable

With 72 Figures

 Springer

Keith E. Hirst, BSc, PhD
School of Mathematics
University of Southampton
Southampton
SO17 1BJ
UK

Cover illustration elements reproduced by kind permission of:
Aptech Systems, Inc., Publishers of the GAUSS Mathematical and Statistical System, 23804 S.E. Kent-Kangley Road, Maple Valley, WA 98038, USA. Tel: (206) 432-7855 Fax (206) 432-7832 email:info@aptech.com URL:www.aptech.com.
American Statistical Association: Chance Vol 8 No 1, 1995 article by KS and KW Heiner 'Tree Rings of the Northern Shawangunks' page 32 fig 2.
Springer-Verlag: Mathematica in Education and Research Vol 4 Issue 3 1995 article by Roman E Maeder, Beatrice Amrhein and Oliver Gloor 'Illustrated Mathematics: Visualization of Mathematical Objects' page 9 fig 11, originally published as a CD ROM 'Illustrated Mathematics' by TELOS: ISBN 0-387-14222-3, German edition by Birkhauser: ISBN 3-7643-5100-4.
Mathematica in Education and Research Vol 4 Issue 3 1995 article by Richard J Gaylord and Kazume Nishidate 'Traffic Engineering with Cellular Automata' page 35 fig 2. Mathematica in Education and Research Vol 5 Issue 2 1996 article by Michael Trott 'The Implicitization of a Trefoil Knot' Page 14.
Mathematica in Education and Research Vol 5 Issue 2 1996 article by Lee de Cola 'Coins, Trees, Bars and Bells: Simulation of the Binomial Process' page 19 fig 3. Mathematica in Education and Research Vol 5 Issue 2 1996 article by Richard Gaylord and Kazume Nishidate 'Contagious Spreading' page 33 fig 1. Mathematica in Education and Research Vol 5 Issue 2 1996 article by Joe Buhler and Stan Wagon 'Secrets of the Madelung Constant' page 50 fig 1.

British Library Cataloguing in Publication Data
Hirst, Keith
    Calculus of one variable. - (Springer undergraduate
    mathematics series)
    1. Functions of real variables
    I. Title
    515.8'3
ISBN-10: 1852339403

Library of Congress Control Number: 2005925984

Springer Undergraduate Mathematics Series ISSN 1615-2085
ISBN-10: 1-85233-940-3   e-ISBN 1-84628-222-5   Printed on acid-free paper
ISBN-13: 978-1-85233-940-1

Printed in the United States of America (HAM)

9 8 7 6 5 4 3 2 1

Springer Science+Business Media
springeronline.com

Preface

The development of the differential calculus was one of the major achievements of seventeenth century European mathematics, originating in the work of Newton, Leibniz and others. Integral calculus can be traced back to the work of Archimedes in the third century B.C. Since its inception, calculus has developed in two main directions. One is the growth of applications and associated techniques, in diverse fields such as physics, engineering, economics, probability and biology. The other direction is that of analytical foundations, where the intuitive and largely geometrical approach is replaced by an emphasis on logic and the development of an axiomatic basis for the real number system whose properties underpin many of the results of calculus. This approach occupied many mathematicians through the eighteenth and nineteenth centuries, culminating in the work of Dedekind and Cantor, leading into twentieth century developments in Analysis and Topology. We can learn much about calculus by studying its history, and a good starting point is the St Andrews' History of Mathematics website www-history.mcs.st-and.ac.uk/history/

This book is designed for beginning university students, both those studying mathematics as a major subject, and those whose main specialism requires the use and understanding of calculus. In the latter case we would expect that lecturers would customise the treatment with applications from the relevant subject area.

The pre-university school mathematics curricula of most European countries all include some calculus, and this book is intended to provide, among other things, a transition between school and university calculus. In some countries such as the U.K., the school curriculum is characterised by an emphasis on techniques and applications, whereas in countries such as France greater attention is given to analytical aspects. In both cases developing understanding of and facility with basic techniques is important, particularly in applied mathematics and statistics. Much of university mathematics is characterised by an emphasis on generality, abstraction and proof. We have included theorems and proofs where they help with understanding the procedures of calculus. We have given definitions of major concepts of calculus, but in relatively informal language. This serves as a link with the more symbolic and logical definitions used in Analysis. Such a development mirrors the transition from calculus to analysis accomplished in the nineteenth century. In this regard we have taken care to give a clear statement of the conditions under which results such as Taylor's Theorem are true, even though we have not included analytical proofs. We have also included a few examples of deductions from definitions, such as the calculation of some of the basic derivatives from first principles. The approach therefore complements texts which do address the analytical foundations, such as Real Analysis, by John M Howie, in this series, referred to as Howie in several places in this book.

The development of the various topics throughout the book is based on the premise that students learn to use and apply techniques through studying well-chosen examples which illustrate general principles, rather than encountering these principles in abstract generality ab initio. Such examples therefore constitute the major component of each chapter, together with the provision of numerous exercises for students to work with. Answers to most of the exercises are included in the book. Full solutions, password protected, are available via the Springer online catalogue. To apply for your password, visit the book page on the Springer online catalogue at www.springeronline.com/1-85233-940-3 or email textbooks@springer-sbm.com.

The mathematical prerequisites for this book are basic algebra and coordinate geometry and the beginnings of differentiation as covered in school. We assume that readers have a reasonable level of facility with algebraic manipulation of straightforward polynomials and rational expressions, the solution of linear and quadratic equations and fractional powers. An appreciation of the basic features of graphs, including the elementary trigonometric, exponential and logarithmic functions is also expected. We also assume a knowledge of co-ordinate geometry associated with straight lines and circles. Most school curricula include basic differentiation with applications to problems involving gradients and maxima and minima. We have based the interface between this book and school mathematics on these topics, revisiting in particular selected parts of algebra, including polynomial division and partial fractions, general properties of functions and their inverses, and some of the elementary results within calculus itself. From this starting point the book develops the differential and integral calculus for functions of one real variable, together with applications from within mathematics, in a systematic and structured form. It covers the subject as commonly featured in first year university courses, and provides a foundation for further work in areas such as differential equations and calculus of more than one variable.

The influence of technology on the teaching and learning of mathematics is the subject of current debate and research. In many countries graphical calculators are prevalent in school mathematics, and many university courses are now integrating the use of computer algebra systems into their courses. We have illustrated the latter in a number of places in the book, by giving MAPLE commands which students with access to this popular, easy to use package can utilise as a basis for exploration. Where an alternative package such as Mathematica or Derive is available it can of course be used instead of MAPLE. In many universities a computer laboratory class is offered as an adjunct to lectures, whereby students can be taught the basic features of such a package. The book is however not designed as a systematic treatise on the use of MAPLE. Readers without access to a computer algebra system will not be disadvantaged, as the text itself is not written so as to be dependent on that facility. Where it is available however it helps to illustrate some concepts, and enables students to explore a much wider range of examples than could be done "by hand". Sample MAPLE worksheets are included on the website referred to above. Instructors are free to use or adapt them for laboratory classes.

I would like to acknowledge the contribution of numerous students at the University of Southampton over many years. Our students are at the core of the drive to communicate mathematics, and I have gained much valuable insight into learning and teaching from them. I would particularly like to thank Jo Bishopp and Linda Walker, who read and commented on the first draft of the book in its entirety, and to Claire Vatcher, who read a later draft and checked the solutions to the exercises. The editorial and technical staff of Springer have been unfailingly helpful. Most of the figures in this book were initially produced using MAPLE, and I would particularly like to thank Aaron Wilson, who improved them significantly. I would also like to thank the publisher's reviewers who supplied many detailed and carefully considered comments which I feel sure have improved the book. In particular they suggested the inclusion of a small amount of material not usually part of an introductory calculus course, which I have included in some chapters by way of added interest and enrichment. Naturally any remaining errors and deficiencies are mine.

Finally I would like to acknowledge my undying gratitude to my dear wife Ann, without whose encouragement and support none of this would be possible.

Keith E Hirst
University of Southampton
June 2005

Contents

1

Functions and Graphs

In this chapter we review some of the basic ideas about functions, and discuss properties of the common families of functions which we encounter in the study of calculus. We have included some of the algebra of functions needed elsewhere in calculus.

The chapter reinforces and extends topics on algebra and functions from pre-university mathematics, and so in some cases we shall simply present a brief revision by way of cementing the links between the contents of this book and readers' previous mathematical studies.

1.1 Functions and Graphs

We consider two familiar graphs, corresponding to the equations $y = x^2$ and $x^2 + y^2 = 1$ respectively, and discuss an important difference between them.

For the first graph, the parabola $y = x^2$, given any value of x we find that there is a unique value of y for which the point with coordinates (x, y) lies on the graph. The second graph is that of the circle $x^2 + y^2 = 1$, with centre the origin and radius 1. In this case, corresponding to any value of x satisfying $-1 < x < 1$, there is more than one value of y for which (x, y) lies on the graph. In the first case, where we have uniqueness, we say that y is a function of x. In the second case the relationship between x and y is not a functional relationship.

Graphically the uniqueness of the value of y corresponds to the fact that

1

the line through the value x parallel to the y-axis does not intersect the graph more than once. This is shown in Figure 1.1 using the parabola and the circle discussed above. It is sometimes referred to as the "vertical line test".

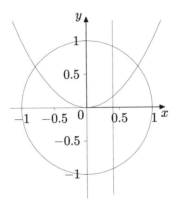

Figure 1.1   The vertical line test

Another way of appreciating this distinction is to consider the function buttons on a calculator. Like the equation for the parabola above, each function button has the property that for each number entered into the calculator (corresponding to x) pressing the button outputs a unique number (corresponding to y). So if for example we enter the number 3.2 and press the x^2 button, we will get the same number every time, namely 10.24.

Readers should be familiar from their previous mathematical studies with the basic properties and graphs of the exponential and logarithmic functions, the trigonometric functions sin, cos and tan, linear functions, and basic powers such as x^2 and x^3, together with reciprocal powers such as x^{-1}.

1.2 Domain and Range

To describe a function f completely we have to specify the set of values of x to be used, as well as the rule or formula which determines the value of y corresponding to each permitted value of x. The set of values of x is called the **domain** of f and the set of all possible values of y which arise is called the **range** of f. This is a very general description, and in practice, throughout this book, a function will be specified by means of a formula, and the domain will be some set of real numbers.

There are occasions where restrictions on the domain arise for other than algebraic reasons. For example a particular formula may describe the speed of a particle at a time t after motion begins. In this case the domain for the formula cannot contain negative values of t for physical reasons. Also there are restrictions on the size and type of number which can be entered into a calculator, and so the domain for each function button will be restricted for electronic reasons. These considerations do not enter into our discussion in this chapter.

What we are normally concerned with is the set of all possible values of x which can legitimately be substituted into the formula for $f(x)$. This is sometimes referred to as the **maximal domain** corresponding to the formula $f(x)$.

In relation to the elementary functions we consider in calculus, the maximal domain is often the set of all real numbers, except when logarithms, square roots etc. are involved, or where algebraic fractions have denominators which are zero at certain values of x. Such restrictions are programmed into calculators. For example if we enter $x = -2$ and then press the square root button, or if we try to divide by zero, the calculator will return an error message.

Example 1.1

Here we have tabulated the domain and range for some basic functions.

Formula	Maximal Domain	Range
$\sin x$	All real x	$-1 \leq y \leq 1$
$\tan x$	$x \neq \frac{(2n+1)\pi}{2}, n = 0, \pm 1, \pm 2 \ldots$	All real y
e^x	All real x	$y > 0$
$\ln x$	$x > 0$	All real y
$\sqrt{x}$	$x \geq 0$	$y \geq 0$

Notice the range for the square root function. We adopt the convention that for any number $x > 0$, $\sqrt{x}$ always denotes the positive number whose square is x. This is the choice programmed into calculators. There are of course two solutions for the equation $y^2 = x$. One is $\sqrt{x}$ and the other one is $-\sqrt{x}$. (Note that 0 has only one square root, namely 0 itself.)

Example 1.2

Determine the maximal domain and corresponding range for the function given by the formula $f(x) = \dfrac{2 - x}{x - 1}$.

The only value of x which cannot legitimately be substituted into the formula is $x = 1$, for which the denominator is zero. The maximal domain is therefore the set of all real numbers except for $x = 1$. The range can be deter-

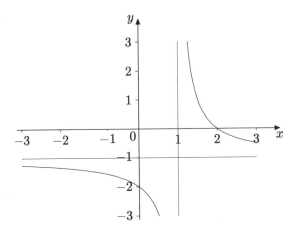

Figure 1.2   Graph of $f(x) = \frac{2-x}{x-1}$

mined from the graph, shown in Figure 1.2, or by rearranging the formula, as follows.

If $y = \dfrac{2-x}{x-1}$ then this equation is equivalent to $x = \dfrac{y+2}{y+1}$, and the only value of y which will not occur is $y = -1$, for which the denominator is zero. Hence the range is the set of all real numbers except $y = -1$.

The horizontal and vertical lines shown in Figure 1.2, which the graph approaches when x and y are increasingly large (positive or negative) are **asymptotes**. They are shown on several graphs in this chapter, and are discussed further in Definition 2.9 in Section 2.3

Example 1.3

Determine the maximal domain and corresponding range for the function given by the formula $f(x) = \ln(\cos x)$.

We know that $\ln t$ is defined only for $t > 0$, and so $\ln(\cos x)$ is defined only when $\cos x > 0$. The maximal domain of f therefore consists of all numbers x for which $\cos x > 0$. We can be more explicit about these values of x by using our knowledge of the cosine function. In particular the graph of the cosine function enables us easily to determine those intervals for which cosine is positive. This occurs when

$$-\frac{5\pi}{2} < x < -\frac{3\pi}{2}; \quad -\frac{\pi}{2} < x < \frac{\pi}{2}; \quad \frac{3\pi}{2} < x < \frac{5\pi}{2}; \quad \text{etc.}$$

There are infinitely many such intervals, and we can write them together in the general form

$$2n\pi - \frac{\pi}{2} < x < 2n\pi + \frac{\pi}{2}; \quad n = 0, \pm 1, \pm 2, \ldots$$

We consider the graph of $\ln(\cos x)$ in Example 1.10.

1.3 Plotting Graphs using MAPLE

A knowledge of the graphs of the elementary functions discussed in this chapter is important, as is some facility with graph sketching "by hand", so that one can deduce some of the important features of a graph related to an algebraic expression. It is essential therefore not to become over-reliant on graph plotting technology. However the use of a graphical calculator or the plotting facilities of a computer algebra system can greatly enrich the understanding of functions and their graphs, and provide a huge range of examples very efficiently. In this section we shall discuss some of the elementary plotting commands in MAPLE. Readers familiar with another package, or a graphical calculator, can replace MAPLE routines where they occur in this book. Readers without access to such technology can simply omit those sections without significant disruption to their progress through calculus.

We shall take as an example the graph in Fig 1.2. This can be plotted (not including the asymptotes) using the following MAPLE command.

```
plot((2-x)/(x-1),x=-3..3,y=-3..3);
```

The syntax of this command is as follows. It begins obviously with the instruction `plot`. We then enter the formula, in which we can use x or some other letter for the variable. After a comma we enter the domain we wish to use for the plot, in this case the interval $-3 \le x \le 3$, the two numbers being separated by two dots as shown. If we omit this we see an error message. It is optional to specify an interval on the y-axis. In this case if we omit it MAPLE will choose something like the interval between -400 and 1000, which will not give a very informative picture of the graph. When we see this result, we can simply edit the original command by entering the interval for y that we want, and we can easily experiment to decide which gives a helpful picture. Finally we see that all MAPLE commands should end with a semi-colon (;) which tells MAPLE that this is the end of the command and that it should be executed.

Readers who are not familiar with this basic MAPLE command can try some familiar functions, such as

```
plot(cos(x),x=-2*Pi..2*Pi);
```

This command has several important features. Firstly it tells us that MAPLE has the cosine function already programmed, and that we must type `cos(x)` with parentheses. Secondly, the symbol * must be used for multiplication. We will see an error message if we omit it; it is one of the errors most people make from time to time. Finally we see that the number π is pre-programmed, denoted in MAPLE by `Pi`, beginning with a capital letter.

The simple `plot` command is used for producing graphs of functions. It cannot therefore be used to draw a circle from an equation such as $x^2 + y^2 = 4$. This equation does not express y as a function of x or vice-versa. Its graph does not satisfy the vertical line test discussed in Section 1.1. It is an example of an implicit equation, for which MAPLE has a separate command. In fact there is a huge variety of MAPLE commands, so that loading them all into the computer's memory at once would be likely to clog up the machine. They are therefore contained in a number of libraries of related routines, which we load only when we need them. So to produce the graph corresponding to an implicit equation we begin the MAPLE session with the command `with(plots);` which accesses plotting commands not included in the basic set. The circle can then be plotted with the command

> `implicitplot(x^2+y^2=4,x=-2..2,y=-2..2,scaling=constrained);`

In this case we must specify the interval on both axes. The command `scaling=constrained` ensures that we see equal scales on the screen. If we omit this then in some cases a circle will appear on the screen shaped like an ellipse.

Sometimes we have to use implicit plotting to obtain a complete picture, for example where roots are involved. So the command

> `plot(root(x,3),x=-2..2);`

will only give the part of the graph of $y = \sqrt[3]{x}$ for which $y \geq 0$. To obtain a complete picture we need to use

> `implicitplot(y^3=x,x=-2..2,y=-2..2);`

MAPLE can plot more than one graph on the same picture, both in the basic `plot` context, and also in commands such as `implicitplot`. For example Figure 1.1 was plotted using

> `implicitplot([x^2+y^2=1, y=x^2,x=0.4], x=-1.1..1.1,`
> `y=-1.3..1.3,scaling=constrained,color=black);`

Finally we note that MAPLE has extensive help facilities, and so if we type `?plot;` we will see a description of the syntax associated with this command, and a selection of illustrative examples, which one can copy into the MAPLE working area to try. (These examples are often more helpful than the syntax description.)

Many of the graphs in this book were plotted using MAPLE, and we shall give examples of MAPLE commands at various places throughout the book.

1.4 Odd and Even Functions

When we are sketching graphs we often use ideas of symmetry, and there are two types of symmetry which are easy to relate to algebraic formulae.

A function is said to be **even** if its graph has reflective symmetry in the y-axis, so that the graph is unchanged by the action of such a reflection. Standard examples are x^2 and $\cos x$. Algebraically this is achieved by replacing x by $-x$, and for an even function the formula resulting from this substitution should be equivalent to the original one. For the examples here we know that $(-x)^2 = x^2$ and $\cos(-x) = \cos x$. This gives rise to the following definition:

Definition 1.4

A function $f(x)$ is an **even function** if $f(-x) = f(x)$ for all x in the domain.

A function is said to be **odd** if its graph is unaltered after reflection in the y-axis followed by reflection in the x-axis. Standard examples are x^3 and $\sin x$. So this time if we replace x by $-x$ then y changes to $-y$. For these examples $(-x)^3 = -x^3$ and $\sin(-x) = -\sin x$. This gives rise to the following definition:

Definition 1.5

A function $f(x)$ is an **odd function** if $f(-x) = -f(x)$ for all x in the domain.

As well as verifying that particular formulae correspond to even or odd functions (or neither), by replacing x by $-x$ and investigating the result, we can prove general results by deduction from the definitions, as in the following examples.

Example 1.6

Show that if a function is both even and odd then it must be zero at every point of its domain.

If f is even then for all x in the domain we have $f(-x) = f(x)$, from the definition. If f is odd then for all x in the domain we have $f(-x) = -f(x)$, again from the definition. From these two equations we deduce that for all x in the domain $f(x) = -f(x)$, and therefore $f(x) = 0$.

Example 1.7

Show that any function can be expressed uniquely as a sum of an even function and an odd function.

We want to find a relationship of the form $f(x) = E(x) + O(x)$, where E is an even function and O is an odd function. If we have such a relationship then replacing x by $-x$ gives

$$f(-x) = E(-x) + O(-x) = E(x) - O(x),$$

using the definitions of even and odd. Adding the two equations and dividing by 2 then gives

$$\frac{f(x) + f(-x)}{2} = E(x).$$

Subtracting gives

$$\frac{f(x) - f(-x)}{2} = O(x).$$

Conversely if we define functions E and O in terms of f by means of these expressions it is easy to verify that E is in fact even and O is odd, as follows.

$$E(-x) = \frac{f(-x) + f(x)}{2} = \frac{f(x) + f(-x)}{2} = E(x).$$

$$O(-x) = \frac{f(-x) - f(x)}{2} = -\frac{f(x) - f(-x)}{2} = -O(x).$$

Example 1.8

We will illustrate the above decomposition using $f(x) = x^2 - 2x - 5$, which is neither even nor odd.

Using the formulae above to work out the even and odd parts gives

$$E(x) = \frac{(x^2 - 2x - 5) + ((-x)^2 - 2(-x) - 5)}{2} = x^2 - 5;$$

$$O(x) = \frac{(x^2 - 2x - 5) - ((-x)^2 - 2(-x) - 5)}{2} = -2x.$$

We can see that $E(x)$ is indeed an even function, $O(x)$ is an odd function, and $f(x) = E(x) + O(x)$.

There are many examples of even and odd functions in this chapter. Graphs of even functions can be seen in Figures 1.4, 1.5, 1.11, 1.13. Graphs of odd functions can be seen in Figures 1.10, 1.12, 1.14, 1.15, 1.17, 1.21, 1.22. In Figures 1.2, 1.6, 1.7 we see examples which are neither even nor odd.

1.5 Composite Functions

The function defined by $f(x) = \ln(\cos x)$ which we discussed in Example 1.3 is built up in stages from functions whose properties we are familiar with, and in that example we were able to determine the domain by considering the properties of ln and cos separately. This method of combining functions is called **composition**, and occurs a good deal in calculus and algebra. There is a special notation, given in the following definition

Definition 1.9

Given two functions f and g, the composition of f with g is defined as follows,

$$f \circ g(x) = f(g(x)).$$

As with Example 1.3 we have to be careful about the domain. Not only must x belong to the domain of g, but $g(x)$ must belong to the domain of f. We read the notation as "f circle g", or "f of g".

On some calculators composition is implemented by using successive function buttons, as follows

$$\text{input } x \quad \overset{\text{press cos}}{\longrightarrow} \quad \cos x \quad \overset{\text{press ln}}{\longrightarrow} \quad \text{output } \ln(\cos x)$$

Clearly we are not limited to two stages in the construction of composite functions. For example we might(!) want to consider the function defined by

$$f(x) = \sqrt{\ln(\tan(e^x))},$$

which is a four-stage composition involving the exponential, tangent, logarithmic and square root functions in succession.

One of the useful applications of composition is in sketching graphs.

Example 1.10

Sketch the graph of $y = \ln(\cos x)$.

We decompose the function by using the intermediate variable t, writing $y = \ln t$ and $t = \cos x$.

The graphs of these two functions are familiar, and we can use them to construct the composite graph, as follows.

In Example 1.3 we determined the domain for $\ln(\cos x)$ and Figure 1.3 confirms that we must have t positive, so the composite graph cannot involve any values of x for which cosine is negative or zero.

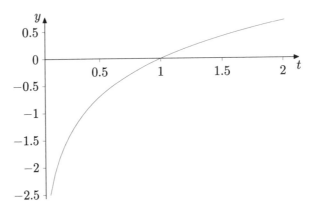

Figure 1.3   Graph of $y = \ln t$

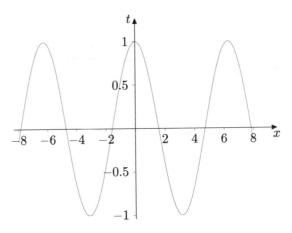

Figure 1.4   Graph of $t = \cos x$

Figure 1.3 shows ln as an increasing function, so that wherever $\cos x$ is increasing, so is $\ln(\cos x)$, and wherever $\cos x$ is decreasing, so is $\ln(\cos x)$. Figure 1.4 reminds us that for all values of x we have $t \leq 1$, and so $\ln t \leq 0$. Finally we see that as t gets smaller, $\ln t$ becomes very large and negative. Figure 1.4 also reminds us that the maximum value of t is 1, and using the increasing property of ln tells us that $\ln(\cos x)$ has a maximum value of $\ln 1 = 0$ at corresponding values of x. We now have all the information we need to construct the composite graph, which is therefore as shown in Figure 1.5.

We can use MAPLE to define and plot functions another way. In connection with Example 1.10, we use the commands

```
g:=x->cos(x);
f:=t->ln(t);
```

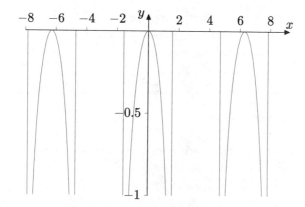

Figure 1.5   Graph of $y = \ln(\cos x)$

to define the cosine and logarithmic functions. We can then define the composition by

```
h:=f(g(x));
```

The following plotting command will then produce a graph similar to Figure 1.5 (without the vertical lines).

```
plot(h,x=-8..8,y=-1..0.2);
```

This exhibits the fact that we have the composition of two functions, in a more explicit way than the single command

```
plot(ln(cos(x)),x=-8..8,y=-1..0.2);
```

1.6 Some Elementary Functions

In this section we provide a brief survey of the elementary functions encountered in calculus. We shall discuss polynomials, rational functions, the modulus function, trigonometric, exponential, logarithmic and hyperbolic functions.

1.6.1 Polynomials

A real polynomial function is a function whose domain is the set of all real numbers, defined by

$$f(x) = a_n x^n + a_{n-1} x^{n-1} + \cdots + a_2 x^2 + a_1 x + a_0 \left(= \sum_{k=0}^{n} a_k x^k \right),$$

where $a_n, \ldots, a_0$ are real constants (the coefficients) and $a_n \neq 0$. The integer n is called the degree of the polynomial, denoted by $\deg(f)$. Polynomials of small degree have particular names which will be familiar to the reader.

degree(n)	0	1	2	3	4	5
	constant	linear	quadratic	cubic	quartic	quintic

A constant polynomial for which $a_0 = 0$ is called the zero polynomial. Its graph coincides with the x-axis.

Polynomials can be added, subtracted and multiplied using the procedures of elementary algebra. In the case of division, if we are given two polynomials it is not always the case that one will divide exactly into the other. Just as with numbers, if we divide a polynomial by a smaller one we would expect to get a quotient and a remainder. This is expressed in the following result.

Theorem 1.11 (The Division Theorem)

Let $f(x), g(x)$ be two polynomials with $\deg(f) \leq \deg(g)$. Then there are unique polynomials $q(x), r(x)$ such that

$$g(x) = q(x)f(x) + r(x),$$

where $\deg(r) < \deg(f)$.

Proof

The proof of this result is straightforward but somewhat tedious. The method of proof follows the same procedure as with any particular example, so it simply describes in general terms the division algorithm, illustrated in Example 1.12.

We shall however prove uniqueness. The method of proof is a common one for uniqueness results. We assume that there are two different possibilities and then deduce that they must be the same after all. In this case suppose that
$g(x) = q_1(x)f(x) + r_1(x)$, where $\deg(r_1) < \deg(f)$, and that
$g(x) = q_2(x)f(x) + r_2(x)$, where $\deg(r_2) < \deg(f)$.
Subtracting and rearranging these two equations gives

$$(q_1(x) - q_2(x)) f(x) = r_2(x) - r_1(x).$$

Now the degree of $r_2(x) - r_1(x)$, the right-hand side, is strictly less than the degree of $f(x)$, because both r_1 and r_2 have degree less than that of f. The left-hand side is a polynomial $(q_1(x) - q_2(x))$ multiplied by $f(x)$ and so the degree of the left-hand side cannot be strictly less than that of f unless $q_1(x) - q_2(x)$ is the zero polynomial. Therefore we must have $q_1 = q_2$, and consequently $r_1 = r_2$, proving uniqueness. $\qquad\square$

The calculations involved in polynomial division may not be familiar to all readers, so we include an example.

Example 1.12

Divide $g(x) = 2x^4 - 3x^3 + 7x^2 - 4x + 1$ by $f(x) = x^2 - 4x + 7$.

The way the algorithm begins is to consider terms of highest degree from g and f respectively, and divide them. So we first divide $2x^4$ by x^2, giving $2x^2$. We then write $g(x) = 2x^2 f(x) + a(x)$, i.e.,

$$2x^4 - 3x^3 + 7x^2 - 4x + 1 = 2x^2(x^2 - 4x + 7) + a(x).$$

We can determine $a(x)$ from this equation, giving

$$a(x) = 5x^3 - 7x^2 - 4x + 1.$$

We repeat the process, this time dividing $a(x)$ by $f(x)$, so again considering the terms of highest degree, we divide $5x^3$ by x^2, giving $5x$. We then need to find $b(x)$ to satisfy

$$a(x) = 5x(x^2 - 4x + 7) + b(x), \quad \text{i.e.,}$$

$$5x^3 - 7x^2 - 4x + 1 = 5x(x^2 - 4x + 7) + b(x).$$

From this equation we find that $b(x) = 13x^2 - 39x + 1$.

Repeating the process again, we divide $13x^2$ by x^2, giving 13, and we then want to find $c(x)$ satisfying

$$b(x) = 13(x^2 - 4x + 7) + c(x), \quad \text{i.e.}$$

$$13x^2 - 39x + 1 = 13(x^2 - 4x + 7) + c(x),$$

and this gives $c(x) = 13x - 90$.

The degree of $c(x)$ is less than that of the divisor $f(x)$ and so the process has finished. We can now assemble all the steps together to give

$$2x^4 - 3x^3 + 7x^2 - 4x + 1 = (2x^2 + 5x + 13)(x^2 - 4x + 7) + 13x - 90.$$

Finally we can divide both sides by $x^2 - 4x + 7$ to obtain

$$\frac{2x^4 - 3x^3 + 7x^2 - 4x + 1}{x^2 - 4x + 7} = 2x^2 + 5x + 13 + \frac{13x - 90}{x^2 - 4x + 7}.$$

The procedure can be set out in the conventional long division format, namely

$$
\begin{array}{r}
2x^2 \quad +5x \quad +13 \\
x^2 - 4x + 7 \overline{)\, 2x^4 \quad -3x^3 \quad +7x^2 \quad -4x \quad +1} \\
2x^4 \quad -8x^3 \quad +14x^2 \\
\overline{\; 5x^3 \quad -7x^2 \quad -4x \quad +1} \\
5x^3 \quad -20x^2 \quad +35x \\
\overline{\; 13x^2 \quad -39x \quad +1} \\
13x^2 \quad -52x \quad +91 \\
\overline{\; 13x \quad -90}
\end{array}
$$

The case when the divisor f is a linear polynomial is particularly important in factorisation, as in the next two theorems.

Theorem 1.13 (The Remainder Theorem)

The remainder on dividing $g(x)$ by $(x - k)$ is $g(k)$, for any real number k.

Proof

From the division theorem we have $g(x) = q(x)(x - k) + r(x)$, where the degree of r is less than 1 (the degree of $(x - k)$). So $r(x)$ is a constant polynomial C, and we therefore have $g(x) = q(x)(x - k) + C$. This is true for all x and so if we substitute $x = k$ we obtain $C = g(k)$, proving the theorem. $\square$

Theorem 1.14 (The Factor Theorem)

If $g(k) = 0$ then $(x - k)$ is a factor of the polynomial $g(x)$ (and vice-versa).

Proof

The result follows from the remainder theorem, because $g(k) = 0$ if and only if $C = 0$, i.e., $g(x) = q(x)(x - k)$. $\square$

When $g(k) = 0$ we say that k is a **root** of the polynomial. This can help us to find linear factors, especially where a polynomial has a root which is a small integer.

Example 1.15

Let $g(x) = x^n - 1$. Then $g(1) = 0$, so $(x - 1)$ is a factor. Polynomial division gives
$$
x^n - 1 = (x - 1)(x^{n-1} + x^{n-2} + \cdots + x^2 + x + 1).
$$

This is a generalisation of the well-known factorisation $x^2 - 1 = (x-1)(x+1)$, and is sufficiently useful to be committed to memory.

Example 1.16

Factorise the cubic polynomial $g(x) = x^3 - x^2 - 5x - 3$.

Factorising quadratics is a familiar topic in school mathematics, and if we have a quadratic such as $x^2 - 2x - 8$ we first look for roots which are integer factors of the constant term -8. The same procedure acts as a starting point for factorising polynomials of higher degree. In this case therefore the natural integers to try as possible roots are the factors of the constant term -3, namely $1, -1, 3, -3$. It is easy to see by substitution that 3 is a root, and so $(x-3)$ is a factor. Polynomial division then gives

$$x^3 - x^2 - 5x - 3 = (x-3)(x^2 + 2x + 1) = (x-3)(x+1)^2.$$

In this kind of factorisation we say that $x = -1$ is a multiple root (in this case a double root because $(x+1)$ appears twice in the factorisation), and that $(x+1)$ is a repeated factor. The general definition is as follows.

Definition 1.17

If the polynomial $g(x)$ can be expressed as $(x-k)^m f(x)$, where $f(k) \neq 0$, then we say that k is a root of g of **multiplicity** m, and that $(x-k)$ is a repeated factor of multiplicity m.

The idea of multiplicity is illustrated by the following two factorisations.

$$x^9 + 3x^8 - 18x^7 - 46x^6 + 129x^5 + 243x^4 - 416x^3 - 504x^2 + 432x + 432$$
$$= (x+1)^2(x+3)^3(x-2)^4.$$

$$x^8 + 5x^7 + 8x^6 + 34x^5 + 69x^4 - 99x^3 - 54x^2 - 324x - 1944$$
$$= (x-2)(x+3)^3(x^2 - x + 6)^2.$$

In the second case the real factorisation can be followed by further factorisation if complex numbers are allowed, giving

$$x^8 + 5x^7 + 8x^6 + 34x^5 + 69x^4 - 99x^3 - 54x^2 - 324x - 1944$$
$$= (x-2)(x+3)^3 \left(x - \frac{1}{2} - \frac{\sqrt{23}}{2}i \right)^2 \left(x - \frac{1}{2} + \frac{\sqrt{23}}{2}i \right)^2.$$

The following theorem gives a method of determining the multiplicity of a root.

Theorem 1.18 (Multiple Roots)

The polynomial $g(x)$ has a root k of multiplicity m if and only if

$$g(k) = g'(k) = g''(k) = \cdots = g^{(m-1)}(k) = 0 \quad \text{and} \quad g^{(m)}(k) \neq 0,$$

where $g', g'', \ldots, g^{(m)}$ denote the successive derivatives of g.

Instead of giving a general proof of this result we shall consider an example.

Example 1.19

Let $g(x) = (x-2)^3(x+1) = x^4 - 5x^3 + 6x^2 + 4x - 8$. Differentiating gives

$$g'(x) = 4x^3 - 15x^2 + 12x + 4; \quad g''(x) = 12x^2 - 30x + 12; \quad g'''(x) = 24x - 30.$$

Now $g(-1) = 0$; $g'(-1) = -27 \neq 0$. This verifies that -1 is a root of multiplicity 1, so it is not repeated. But

$$g(2) = 0; \quad g'(2) = 0; \quad g''(2) = 0; \quad g'''(2) = 18 \neq 0,$$

which verifies that 2 is a root of multiplicity 3.

The theory behind factorisation relates to the Fundamental Theorem of Algebra, proved by K.F. Gauss in 1800. This result implies that any polynomial with real coefficients will factorise into a product of real linear and quadratic factors, some of which may be repeated factors.

The problem with factorisation is that there is no general algorithm, so that the theorem simply says that the factors exist, without telling us how to find them.

Finding factors is equivalent to finding the roots of an equation. As far as quadratics are concerned, the Arabic mathematician Al-Khwarizmi (around 800 A.D.) gave what is in effect the familiar formula for solving quadratics.

Methods for solving cubic equations and quartic equations were developed during the sixteenth century in Italy. The chief mathematicians involved were probably Scipione dal Ferro, Girolamo Cardano and Nicolo of Brescia (also known as Tartaglia). Naturally there were attempts subsequently to develop formulae for solving equations of higher degree, and so it was something of a surprise when in 1799 Paolo Ruffini claimed to offer a proof that there could be no formula for solving equations of degree five. The mathematician usually credited with the first complete proof of this result is the Norwegian Niels Henrik Abel, in 1824.

There is more information about this part of the history of mathematics on the St. Andrews' website `www-history.mcs.st-and.ac.uk/history/`

1.6.2 Rational Functions

A rational function is a quotient of two polynomials. The following expressions represent rational functions.

$$\frac{2x^2 - 5x + 6}{x - 2}; \quad \frac{x + 4}{2x - 5}; \quad \frac{3x - 3}{x^4 - 3x^2 + x - 17}; \quad \frac{x^3 + 1}{x^3 - 1}.$$

The following quotients do **not** define rational functions.

$$\frac{x^2 + 3}{\cos^2(x + 2) - 7x}; \quad \frac{x^3 - 3x^2 - 5}{\sqrt{2x + 3}}; \quad \frac{\ln x - \tan x}{e^{2x-4} + \sec x}.$$

One important aspect of rational functions is their decomposition into simpler rational functions known as partial fractions. This is especially important in integration, and is discussed in detail in Chapter 10 where we consider integration of rational functions.

The form of the graph of a rational function $y = f(x)$ depends on a number of properties, as follows.

1. If the degree of the numerator is greater than that of the denominator then when x becomes large (positive or negative) so does y.

2. If the degree of the numerator is equal to that of the denominator then there will be asymptotes parallel to the x-axis as x becomes large (positive or negative).

3. If the degree of the numerator is less than that of the denominator then the x-axis itself is an asymptote as x becomes large (positive or negative).

4. If the denominator has no real roots then the graph will be a continuous curve.

5. If the denominator does have real roots there will be asymptotes parallel to the y-axis corresponding to each of the real roots.

We shall discuss some of these in more detail in Chapter 2 when we consider limits of functions. For the moment we shall simply show some examples. Cases 2. and 5. together are illustrated in Figure 1.2. Cases 1. and 4. together are shown in Figure 1.6. Cases 3. and 5. together are shown in Figure 1.7.

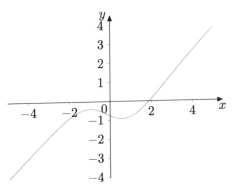

Figure 1.6   Graph of $\frac{x^3-2x-3}{x^2+4}$

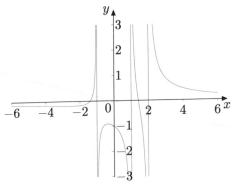

Figure 1.7   Graph of $\frac{x^2-2}{x^3-2x^2-x+2}$

1.6.3 The Modulus Function

The modulus or absolute value function is important in many parts of mathematics, and so familiarity with graphs and algebraic manipulations involving modulus is useful. The modulus function is defined as follows.

Definition 1.20

$$f(x) = \begin{cases} x & \text{if } x \geq 0, \\ -x & \text{if } x < 0. \end{cases}$$

In Figure 1.8 we have plotted $y = |x|, y = |x - 2|, y = |x| - 1$. The modulus function itself is the graph passing through the origin, and the last of the three is shown using MAPLE's point style. Notice how the modulus function has been shifted in each case. Note than MAPLE uses abs(x) for the modulus function.

On the number line the modulus of a number represents its distance from zero, without regard to direction, so it is always a non-negative number. The expression $|x - a|$ measures the distance of x from a, again without regard to direction. We can use this interpretation in solving equations and inequalities involving the modulus function.

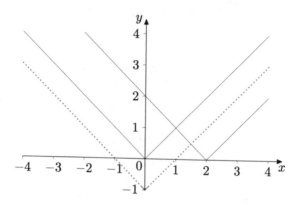

Figure 1.8   Graphs of modulus functions

Example 1.21

Solve the equation $|x + 4| = 3$.

Algebraically, $|x + 4| = 3$ means that $x + 4 = \pm 3$, so the equation has two solutions, $x = -1, x = -7$. Geometrically we are looking for numbers x whose distance from -4 is 3, giving $x = -1, x = -7$ as before. MAPLE can deal with this by using solve(abs(x+4)=3,x);

Example 1.22

Solve the inequality $|x^2 - 2x - 5| \le 3$.

Geometrically this means that the distance of $x^2 - 2x - 5$ from zero is less than or equal to 3, i.e., $-3 \le x^2 - 2x - 5 \le 3$. A good way to solve this is to use a combination of algebra and geometry. Figure 1.9 shows the graphs of $y = x^2 - 2x - 5$ (as the parabola), $y = 3$, $y = -3$ and $y = |x^2 - 2x - 5|$. Where the parabola is positive its graph is the same as that of its modulus. Where the parabola is negative we reflect that portion in the x-axis to obtain the graph of its modulus. So we are looking for those values of x for which the quadratic lies between the horizontal lines $y = 3$ and $y = -3$, or equivalently for which the modulus graph lies below the line $y = 3$.

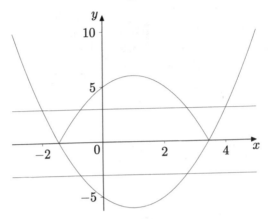

Figure 1.9 Graph for solving $|x^2 - 2x - 5| \leq 3$

We can now see that we want the set of values of x in the two intervals determined by the solutions of the quadratic equations $x^2 - 2x - 5 = 3$ and $x^2 - 2x - 5 = -3$, namely $x = -2, 4, 1 - \sqrt{3}, 1 + \sqrt{3}$. So the solution of the inequality is the set of all numbers x satisfying $-2 \leq x \leq 1 - \sqrt{3}$ or $1 + \sqrt{3} \leq x \leq 4$. MAPLE can solve inequalities as well as equations, in this case using `solve(abs(x^2-2x-5)<=3,x);`

1.6.4 Trigonometric Functions

Readers should be familiar with some of the properties of the three basic trigonometric functions, sine, cosine and tangent, and their graphs. Each of these functions is periodic, so for example $\cos(x + 2\pi) = \cos x$ for all x. Perhaps less familiar are the reciprocals of these three functions.

secant - the reciprocal of cosine: $\sec = \frac{1}{\cos}$;

cosecant - the reciprocal of sine: $\csc = \frac{1}{\sin}$;

cotangent - the reciprocal of tangent: $\cot = \frac{1}{\tan}$.

Note that cos, sin, sec and cosec all have period 2π, whereas tan and cot have period π.

The graphs of the three reciprocal functions can be constructed from the basic functions using the fact that $\frac{1}{x}$ is a decreasing function. So if $\cos x$ is decreasing over some interval then $\sec x$ will be increasing, and vice-versa. Also when $\cos x$ tends to zero through positive values then its reciprocal will tend to infinity, so the graph will have a vertical asymptote (for example when $x = \pi/2$). Figures 1.10, 1.11, 1.12 show the three basic trigonometric functions, and their reciprocals, including asymptotes. In each case we have plotted the graphs over two complete periods.

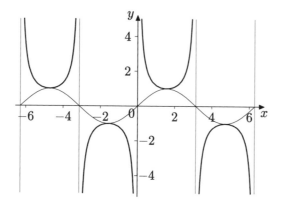

Figure 1.10   Graphs of sine and cosecant

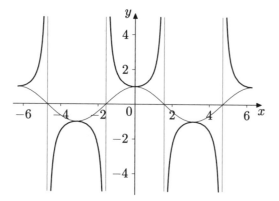

Figure 1.11   Graphs of cosine and secant

We can derive trigonometric identities for these three reciprocal functions from those of the three basic functions, as in the following example. They do not need separate proofs.

Example 1.23

(a) Starting with $\cos^2 x + \sin^2 x = 1$, we can divide both sides by $\sin^2 x$ to give

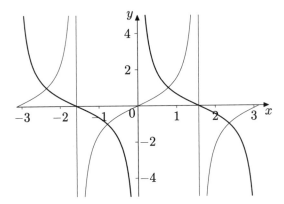

Figure 1.12   Graphs of tangent and cotangent

$\cot^2 x + 1 = \operatorname{cosec}^2 x$, and by $\cos^2 x$ to obtain $1 + \tan^2 x = \sec^2 x$.

(b) Using the addition formulae for sine and cosine we deduce that

$$\cot(x+y) = \frac{\cos(x+y)}{\sin(x+y)} = \frac{\cos x \cos y - \sin x \sin y}{\sin x \cos y + \cos x \sin y} = \frac{\cot x \cot y - 1}{\cot x + \cot y}.$$

(The last step involves dividing numerator and denominator by $\sin x \sin y$.)

What this example demonstrates is the principle that we need only remember a very small number of identities to be able to derive all the others, as in the following example. We shall utilise this principle when we consider other sets of functions.

Example 1.24

We shall make some deductions from the four addition formulae:

$$\cos(x+y) \;=\; \cos x \cos y - \sin x \sin y; \qquad (1.1)$$
$$\sin(x+y) \;=\; \sin x \cos y + \cos x \sin y; \qquad (1.2)$$
$$\cos(x-y) \;=\; \cos x \cos y + \sin x \sin y; \qquad (1.3)$$
$$\sin(x-y) \;=\; \sin x \cos y - \cos x \sin y. \qquad (1.4)$$

In fact if we use the fact that cosine is an even function and that sine is an odd function, as their graphs indicate, then we can deduce (1.3) from (1.1) and (1.4) from (1.2), thereby reducing the number of formulae needed to two. Now putting $y = x$ in (1.1) and (1.2) gives the two double angle formulae

$$\cos 2x \;=\; \cos^2 x - \sin^2 x;$$
$$\sin 2x \;=\; 2 \sin x \cos x.$$

Dividing these two enables us to deduce double angle formulae for tangent and cotangent.

Formulae such as that for $\cos 3x$ are derived by using $\cos 3x = \cos(x + 2x)$ and then utilising the addition and double angle formulae.

Adding (1.1) and (1.3) gives

$$\cos(x + y) + \cos(x - y) = 2\cos x \cos y,$$

which converts a product of trigonometric functions into a sum. This is useful in integration. Similarly we can find formulae for $\sin x \sin y$ and $\sin x \cos y$.

A comprehensive list of trigonometric identities is included for reference in Section 1.6.7.

1.6.5 Exponential and Logarithmic Functions

An acquaintance with the exponential and logarithmic functions forms part of most pre-university mathematics courses, so in this section, as with the trigonometric functions, we shall briefly revise basic properties and give a few examples of interest.

Later in this chapter we shall be discussing inverse functions, and in that connection we note here the important fact that the exponential and logarithmic functions are inverses of one another. This is embodied in the important relationship:

$$y = \exp(x) \text{ if and only if } x = \ln y. \tag{1.5}$$

We can deduce two useful relationships from this. Firstly by eliminating y we obtain $x = \ln(\exp(x))$, and secondly eliminating x gives $y = \exp(\ln y)$.

Because of equation (1.5) one can adopt two equivalent theoretical approaches to defining the exponential and logarithmic functions.

One can define the exponential function from scratch (as the sum of a certain infinite series, or as a function which is its own derivative, for example), establish its properties, and define the logarithmic function as the inverse of the exponential function. We can then deduce the properties of the logarithmic function from those of the exponential, illustrated in Example 1.26.

On the other hand one could define the logarithmic function $\ln(x)$ from scratch (as the integral $\int_1^x (1/t)\, dt$ for example), establish its properties, define the exponential function as the inverse of the logarithmic function, and finally deduce the properties of the exponential function from those of the logarithm, as in Example 1.25. This approach is exemplified in Howie Chapter 6. In that chapter he proves the basic law of logarithms given as equation 1.9 below. We can then obtain the fundamental property of the exponential function as in the following example.

Example 1.25

Deduce from equation 1.9 below that

$$\exp(x)\exp(y) = \exp(x+y).$$

Beginning from the left-hand side and using equation 1.9 gives

$$\ln(\exp(x)\exp(y)) = \ln(\exp(x)) + \ln(\exp(y)) = x + y,$$

since ln and exp are inverses. Using this fact then gives

$$\exp(x)\exp(y) = \exp(x+y),$$

as required.

We then denote the number $\exp(1)$ by e. In fact $e \approx 2.71828$ correct to five decimal places. If you want to see the first 100 places MAPLE will tell you, using `evalf[100](exp(1));` The property of exponentials established in Example 1.25 tells us that

$$\exp(2) = \exp(1+1) = \exp(1)\exp(1) = e^2,$$

and we can continue this procedure to show that for all positive integers n, $\exp(n) = e^n$.

We also have $\exp\left(\frac{1}{2}\right)\exp\left(\frac{1}{2}\right) = \exp(1) = e$, so $\exp\left(\frac{1}{2}\right) = \sqrt{e} = e^{\frac{1}{2}}$. We can extend this to show that for any rational number $\frac{m}{n}$, $\exp\left(\frac{m}{n}\right) = e^{\frac{m}{n}}$. Finally we define e^x as $\exp(x)$ for any real number x, and the property of the exponential function in Example 1.25 becomes the law of indices in equation 1.6 below. Both notations, e^x and $\exp(x)$, are in common use for the exponential function.

The basic properties of these functions are listed as the laws of indices and the laws of logarithms

$$
\begin{array}{rcll}
e^{x+y} &=& e^x e^y & (1.6) \\
e^{-x} &=& 1/e^x & (1.7) \\
(e^x)^y &=& e^{xy} & (1.8) \\
\ln(xy) &=& \ln x + \ln y & (1.9) \\
\ln(1/x) &=& -\ln x & (1.10) \\
\ln(x^y) &=& y\ln x & (1.11)
\end{array}
$$

Example 1.26

Deduce equation (1.11) above from the properties of the exponential function.

In the following chain of reasoning each line is equivalent to the previous one.

$$
\begin{aligned}
z &= \ln(x^y) \\
e^z &= x^y \quad \text{(inverse function property (1.5))} \\
(e^z)^{(1/y)} &= (x^y)^{1/y} \quad \text{(both sides raised to power } 1/y) \\
e^{(z/y)} &= x^{(y/y)} \quad \text{(law of indices (1.8))} \\
e^{(z/y)} &= x \quad (x^1 = x) \\
\frac{z}{y} &= \ln x \quad \text{(inverse function property (1.5))} \\
z &= y \ln x
\end{aligned}
$$

Most calculators have an x^y button, or something equivalent, and one could use it to obtain a numerical value for $\sqrt{3}^{\sqrt{2}}$. But what does an expression like this mean, and how can it be evaluated? After all we can't imagine $\sqrt{3}$ multiplied by itself $\sqrt{2}$ times! What we would like to do is to give a definition for a^x which is consistent with the laws of indices (and therefore logarithms). If we make this assumption then we have

$y = a^x$ if and only if $\ln y = \ln(a^x) = x \ln a$, equivalent to $y = e^{x \ln a}$.

Definition 1.27

$$
a^x = e^{x \ln a} \quad (a > 0).
$$

Using this definition we now have

$$
\sqrt{3}^{\sqrt{2}} = e^{\sqrt{2} \ln \sqrt{3}} = e^{\sqrt{2} \ln(3^{1/2})} = e^{\sqrt{2}(\ln 3)/2}.
$$

Working this out on a calculator using just the exponential and ln buttons gives the same result as using the x^y button, verifying the relationship (whilst noting that the calculator returns only an approximation to these numbers, accurate to a certain number of decimal places).

As well as the ln button calculators have a log button, which refers to the inverse of 10^x, and in fact some calculators use the label 10^x. In fact now that we have a definition of a^x for $a > 0$, we can define a general logarithm $\log_a x$ as the inverse of a^x. We read it as "the logarithm of x to base a". Such logarithms are largely of theoretical interest, although $\log_2$ is sometimes used in computing. General logarithms obey laws analogous to those for ln, with appropriate restrictions on domains, namely

$$
\begin{aligned}
\log_a xy &= \log_a x + \log_a y; \\
\log_a(x/y) &= \log_a x - \log_a y; \\
\log_a x^y &= y \log_a x.
\end{aligned}
$$

Example 1.28

The laws of logarithms can be used to simplify expressions, for example

$$\log_2 32 = \log_2 2^5 = 5 \log_2 2 = 5 \quad (\log_a a = 1);$$
$$\log_4 32 = \log_4 4^{5/2} = \tfrac{5}{2};$$
$$\log_4 48 = \log_4(16 \times 3) = \log_4 16 + \log_4 3 = \log_4 4^2 + \log_4 3 = 2 + \log_4 3.$$

Before the advent of electronic calculators $\log_{10}$ was very important. There were tables used in school of $\log_{10}$ and its inverse (often called "antilogarithms") and generations of schoolchildren had to learn to use these tables to perform arithmetic calculations. These tables were "four figure tables", given to four decimal places of accuracy. Historically logarithm tables and trigonometric tables were developed for calculations in astronomy, and books of seven figure tables were published.

1.6.6 Hyperbolic Functions

We showed in Example 1.7 that any function can be expressed as a sum of an even function and an odd function. We obtained the formulae

$$\frac{f(x) + f(-x)}{2}; \qquad \frac{f(x) - f(-x)}{2}$$

for the even and odd components.

In the case where $f(x)$ is the exponential function we obtain a pair of functions called hyperbolic functions, defined by

$$\cosh x = \frac{e^x + e^{-x}}{2}; \qquad \sinh x = \frac{e^x - e^{-x}}{2},$$

with graphs as shown in Figures 1.13 and 1.14. Note that cosh is an even function, whereas sinh is odd.

Just as with trigonometric functions we can define corresponding hyperbolic functions

$$\tanh x = \frac{\sinh x}{\cosh x}; \quad \coth x = \frac{\cosh x}{\sinh x}; \quad \operatorname{sech} x = \frac{1}{\cosh x}; \quad \operatorname{cosech} x = \frac{1}{\sinh x}.$$

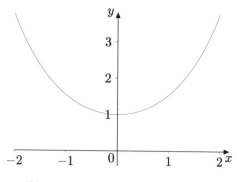

Figure 1.13   Graph for cosh

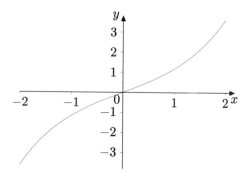

Figure 1.14   Graph for sinh

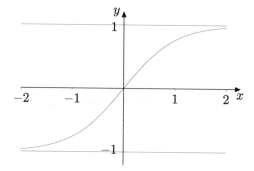

Figure 1.15   Graph for tanh

The graph of $\tanh x$ is shown in Figure 1.15.

There are many identities involving hyperbolic functions, which have both similarities and differences compared with those for trigonometric functions. These can be proved using the definitions of the hyperbolic functions and the laws of indices for exponentials.

Example 1.29

We prove the basic identity $\cosh^2 x - \sinh^2 x = 1$.

$$
\begin{aligned}
\cosh^2 x - \sinh^2 x &= \left(\frac{e^x + e^{-x}}{2}\right)^2 - \left(\frac{e^x - e^{-x}}{2}\right)^2 \\
&= \frac{e^{2x} + 2 + e^{-2x} - (e^{2x} - 2 + e^{-2x})}{4} = 1.
\end{aligned}
$$

Example 1.30

We shall prove an addition formula for sinh

$$
\begin{aligned}
&\sinh x \cosh y + \cosh x \sinh y \\
&= \frac{e^x - e^{-x}}{2} \cdot \frac{e^y + e^{-y}}{2} + \frac{e^x + e^{-x}}{2} \cdot \frac{e^y - e^{-y}}{2} \\
&= \frac{(e^x - e^{-x})(e^y + e^{-y}) + (e^x + e^{-x})(e^y - e^{-y})}{4} \\
&= \frac{2e^x e^y - 2e^{-x} e^{-y}}{4} \\
&= \frac{e^{x+y} - e^{-(x+y)}}{2} = \sinh(x + y).
\end{aligned}
$$

Other identities can be derived in a similar way to the two above, and are listed for reference in Section 1.6.7.

One reason for the name "hyperbolic" is that these functions can be used to parameterise hyperbolae, in the same way that trigonometric functions (sometimes called the circular functions) can be used to parameterise circles (and ellipses). If we consider the cartesian equation of a hyperbola,

$$
\frac{x^2}{a^2} - \frac{y^2}{b^2} = 1,
$$

then putting $x = a \cosh t$, $y = b \sinh t$ and using the identity in Example 1.29 gives

$$
\frac{(a \cosh t)^2}{a^2} - \frac{(b \sinh t)^2}{b^2} = \cosh^2 t - \sinh^2 t = 1.
$$

1.6.7 Trigonometric and Hyperbolic Identities

This section contains a list of identities, for reference. Readers should try to derive those which have not been dealt with earlier, as we did for instance in Example 1.23.

Trigonometric Identities

$$\sin(x + y) = \sin x \cos y + \cos x \sin y \qquad \cos^2 x + \sin^2 x = 1$$

$$\sin(x - y) = \sin x \cos y - \cos x \sin y \qquad 1 + \tan^2 x = \sec^2 x$$

$$\cos(x + y) = \cos x \cos y - \sin x \sin y \qquad \cot^2 x + 1 = \operatorname{cosec}^2 x$$

$$\cos(x - y) = \cos x \cos y + \sin x \sin y$$

$$\tan(x + y) = \frac{\tan x + \tan y}{1 - \tan x \tan y} \qquad \tan(x - y) = \frac{\tan x - \tan y}{1 + \tan x \tan y}$$

$$2 \sin x \cos y = \sin(x + y) + \sin(x - y) \qquad \sin(-x) = -\sin x$$

$$2 \cos x \sin y = \sin(x + y) - \sin(x - y) \qquad \cos(-x) = \cos x$$

$$2 \cos x \cos y = \cos(x + y) + \cos(x - y) \qquad \sin 2x = 2 \sin x \cos x$$

$$2 \sin x \sin y = \cos(x - y) - \cos(x + y) \qquad \tan 2x = \frac{2 \tan x}{1 - \tan^2 x}$$

$$\cos 2x = \cos^2 x - \sin^2 x = 2 \cos^2 x - 1 = 1 + 2 \sin^2 x$$

$$\sin^2\left(\frac{x}{2}\right) = \frac{1 - \cos x}{2} \qquad\qquad \cos^2\left(\frac{x}{2}\right) = \frac{1 + \cos x}{2}$$

Hyperbolic Identities

$$\sinh(x + y) = \sinh x \cosh y + \cosh x \sinh y \qquad \cosh^2 x - \sinh^2 x = 1$$

$$\sinh(x - y) = \sinh x \cosh y - \cosh x \sinh y \qquad 1 - \tanh^2 x = \cosh^2 x$$

$$\cosh(x + y) = \cosh x \cosh y + \sinh x \sinh y \qquad \coth^2 x - 1 = \operatorname{cosech}^2 x$$

$$\cosh(x - y) = \cosh x \cosh y - \sinh x \sinh y$$

$$\tanh(x + y) = \frac{\tanh x + \tanh y}{1 + \tanh x \tanh y} \qquad \sinh 2x = 2 \sinh x \cosh x$$

$$\tanh(x - y) = \frac{\tanh x - \tanh y}{1 - \tanh x \tanh y} \qquad \tanh 2x = \frac{2 \tanh x}{1 + \tanh^2 x}$$

$$\cosh 2x = \cosh^2 x + \sinh^2 x = 2 \cosh^2 x - 1 = 1 - 2 \sinh^2 x$$

$$\sinh(-x) = -\sinh x \qquad\qquad \cosh(-x) = \cosh x$$

$$\sinh^2\left(\frac{x}{2}\right) = \frac{1 - \cosh x}{2} \qquad\qquad \cosh^2\left(\frac{x}{2}\right) = \frac{1 + \cosh x}{2}$$

1.7 Inverse Functions

The equation $y = f(x)$ gives y in terms of x. To solve it for x we need to rearrange it to get x in terms of y. This occurs in many problems where we need to change the subject of a formula. For example consider the formula for the volume of a cylinder $V = \pi r^2 h$. If we are given the values of r and h

then of course we can calculate V. However if we are given V and r we need to rearrange the formula to find h, and this can be done, to give $h = V/\pi r^2$. Here the solution for h is unique, but with an equation such as $y = x^2$ the solution for x is not unique, and so the equation does not determine x as a function of y. On the other hand, starting with $y = x^3$ does give the unique solution $x = \sqrt[3]{y}$, expressing x as a function of y. It is this distinction and its consequences that we shall be exploring in this section. When each value of y in the range of a function given by $y = f(x)$ is associated with a unique value of x in the domain, as with the cube function, we say that the function is one-to-one (1-1). Graphically this means that a function is 1-1 if and only if each line parallel to the x-axis meets the graph in at most one point. This property is sometimes referred to as the "horizontal line test". It is illustrated in Fig 1.16, using the square function and cube function discussed above, where we have shown the horizontal line in three possible positions.

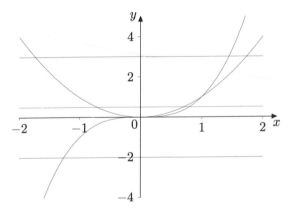

Figure 1.16   The horizontal line test

We can give a symbolic definition as follows.

Definition 1.31

The function $f(x)$ is said to be **one-to-one (1-1)** if $f(a) = f(b)$ implies that $a = b$ for all a, b in the domain of f.

Example 1.32

Prove algebraically that the function defined by $y = x^3$ is 1-1.

If $a^3 = b^3$ then $a^3 - b^3 = 0$, and so $(a - b)(a^2 + ab + b^2) = 0$. Completing the square gives $a^2 + ab + b^2 = \left(a + \frac{1}{2}b\right)^2 + \frac{3}{4}b^2$. This is a sum of squares and so cannot be zero unless both terms are zero, which occurs only if $a = b = 0$. Therefore we must have $a - b = 0$, i.e., $a = b$.

Definition 1.33

If a function f is 1-1 then it has an inverse function, denoted by f^{-1}, where for each x belonging to the range of f, $f^{-1}(x)$ is defined to be the unique value of y in the domain of f satisfying $x = f(y)$.

This definition contains the important relationship that $y = f^{-1}(x)$ and $x = f(y)$ are equivalent. Notice also that the range of f is the same as the domain of f^{-1} and vice-versa.

We met an example of this relationship at the beginning of the discussion of exponential and logarithmic functions in Section 1.6.5.

Notice that when $f(x) = x^3$ we have $f^{-1}(f(x)) = \sqrt[3]{x^3} = x$ and $f(f^{-1}(x)) = (\sqrt[3]{x})^3 = x$. This exemplifies a general property - the inverse simply takes you back to where you started from. It can be verified on a calculator if it has a cube and cube root button, and also with the exponential and ln button.

Since $y = f^{-1}(x)$ and $x = f(y)$ are equivalent their graphs are the same. For example the graphs of $y = \ln x$ and $x = e^y$ are identical. To obtain the graph of $x = f(y)$ from that of $y = f(x)$ we clearly have to interchange x and y. Geometrically this means that every point (x, y) on one graph is transformed to the point (y, x) on the other. This is achieved by means of reflection in the line $y = x$. This is shown in Figure 1.17 which contains the graphs of $y = x^3$ (equivalent to $x = \sqrt[3]{y}$) and its inverse $y = \sqrt[3]{x}$ (equivalent to $x = y^3$). The line $y = x$ is also shown, making clear the reflection property. Figure 1.18 shows the same relationship between the exponential and logarithmic functions.

Quadratic functions are generally not 1-1 unless their domain is restricted. For example using the square and square root operations gives

$$\sqrt{(-3)^2} = \sqrt{9} = 3,$$

(recalling the convention about positive square roots) so that the composite operation (squaring and then square rooting) does not take us back to where we started (the number -3). In the next example we show how to restrict the domain and thereby produce a proper inverse function.

Example 1.34

Let $y = x^2 + 2x = (x + 1)^2 - 1$. This does not define a 1-1 function if we take the domain to be the set of all real numbers, for example if $y = 3$ then x could

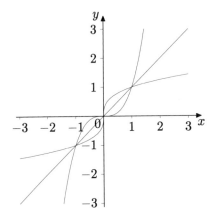

Figure 1.17 $y = x^3$ and its inverse

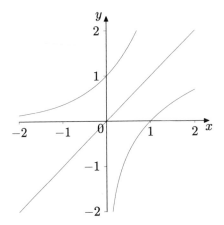

Figure 1.18 Exponential and logarithm as inverses

be either 1 or -3. However if we restrict the domain then we can obtain 1-1 functions. The two functions defined below are both 1-1.

$$f(x) = x^2 + 2x, \quad (x \leq -1);$$
$$g(x) = x^2 + 2x, \quad (x \geq -1).$$

It is important to understand that f and g are different functions. Although the formula used is the same, the domains are different, and their graphs are distinct. In Figure 1.19 the graph of $f(x)$ is shown dotted, and $g(x)$ solid. Now if $y = (x+1)^2 - 1$ then $x = -1 \pm \sqrt{y+1}$. To see which choice of square root corresponds to f and which to g we note that for f we must have $x \leq -1$, and so $y = f(x)$ is equivalent to $x = -1 - \sqrt{y+1}$. For g we have $x \geq -1$ and so $y = g(x)$ is equivalent to $x = -1 + \sqrt{y+1}$. The respective formulae for the

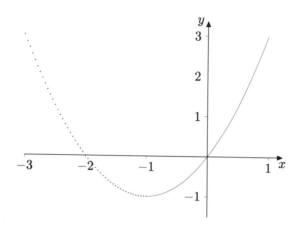

Figure 1.19   Graph for Example 1.34

inverse functions are therefore $f^{-1}(x) = -1 - \sqrt{x-1}$, $g^{-1}(x) = -1 + \sqrt{x+1}$.
One can also draw these conclusions by considering graphs, and readers should
do this.

1.7.1 Increasing and Decreasing Functions

Definition 1.35

A function f with domain D is said to be an **increasing function** if for all a
and b in the domain D, $a \le b$ implies that $f(a) \le f(b)$.
A function f with domain D is said to be a **strictly increasing function** if
for all a and b in the domain D, $a < b$ implies that $f(a) < f(b)$. Notice that
we have strict inequality signs here.
A function f with domain D is said to be a **decreasing function** if for all a
and b in the domain D, $a \le b$ implies that $f(a) \ge f(b)$.
A function f with domain D is said to be a **strictly decreasing function** if
for all a and b in the domain D, $a < b$ implies that $f(a) > f(b)$.
A function f with domain D is said to be **monotonic** if it is either an increasing
function or a decreasing function.

Increasing and decreasing functions have very similar properties, in the sense
that if the graph of an increasing function is reflected in the x-axis it becomes a
decreasing function, and vice-versa. This is why it is often convenient to group
them together as monotonic functions. Examples of increasing functions are

the cube and cube root functions (Figure 1.17), tanh (Figure 1.15), and the exponential and logarithmic functions (Figure 1.18).

As well as discussing these properties graphically we can also verify them algebraically.

Example 1.36

Show that the function e^{-x} is strictly decreasing.

$e^{-a} > e^{-b}$ if and only if $e^{-a}/e^{-b} > 1$, i.e., $e^{b-a} > 1$, using laws of indices. Because $e > 1$, $e^{b-a} > 1$ occurs if and only if $b - a > 0$, i.e., $a < b$. So e^{-x} is strictly decreasing.

The examples given above of strictly increasing functions all have inverses, and this true in general.

Example 1.37

Show that every strictly increasing f function has an inverse.

We need to show that f is 1-1. If $a \neq b$ then either $a < b$ or $b < a$. Therefore, because f is strictly increasing, either $f(a) < f(b)$ or $f(b) < f(a)$, so that $f(a) \neq f(b)$. Therefore if $f(a) = f(b)$ we must have $a = b$.

The same result is true for a strictly decreasing function, by a very similar argument. In other words any strictly monotonic function has an inverse.

If we write down a formula for a function which has an inverse it will often be monotonic. This is because in most cases such a function will have a continuously drawn graph. Proving that function with a continuous graph which has an inverse must be monotonic is outside the scope of this book, and would be encountered in a study of Real Analysis. In fact without the stipulation of a continuous graph the result is not true, as Example 1.38 shows.

Example 1.38

Plot the graph of the following function, with domain given by $0 \leq x \leq 2$, and explain why it has an inverse.

$$f(x) = \begin{cases} x^2 & \text{if } 0 \leq x < 1, \\ 4 - x & \text{if } 1 \leq x \leq 2. \end{cases}$$

We can see clearly from the Figure 1.20 that the function satisfies the horizontal line test and is therefore 1-1. So the function has an inverse.

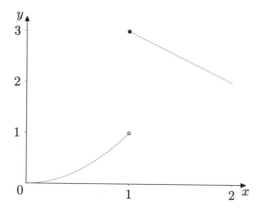

Figure 1.20   A 1-1 non-monotonic function

A somewhat more bizarre 1-1 non-monotonic function with domain given by $0 < x < 1$, can be defined as follows.

$$f(x) = \begin{cases} x & \text{if } x \text{ is a rational number,} \\ 1 - x & \text{if } x \text{ is an irrational number.} \end{cases}$$

Such functions are really outside the scope of this book. They are often used to provide examples in Real Analysis.

1.7.2 Inverse Trigonometric Functions

Looking at the graphs of any of the trigonometric functions tells us that none of them is 1-1. In each case corresponding to a given value of y in the range there are infinitely many values of x in the domain, because of periodicity. This is exemplified in the table of standard results below.

$y = \sin x$	
y	x
0	$n\pi : n = 0, \pm1, \pm2, \cdots$
1	$\frac{\pi}{2} + 2n\pi : n = 0, \pm1, \pm2, \cdots$
-1	$-\frac{\pi}{2} + 2n\pi : n = 0, \pm1, \pm2, \cdots$
$\frac{1}{2}$	$\frac{\pi}{6} + 2n\pi : n = 0, \pm1, \pm2, \cdots$
$-\frac{1}{2}$	$-\frac{\pi}{6} + 2n\pi : n = 0, \pm1, \pm2, \cdots$
$\frac{\sqrt{3}}{2}$	$\frac{\pi}{3} + 2n\pi : n = 0, \pm1, \pm2, \cdots$
$-\frac{\sqrt{3}}{2}$	$-\frac{\pi}{3} + 2n\pi : n = 0, \pm1, \pm2, \cdots$

However, calculators have buttons which calculate values for inverse trigonometric functions. So what is happening? We can try some calculations. From school mathematics you should be familiar with the trigonometrical function values for standard angles (often expressed in degrees, although it is important to become familiar with radians). With the calculator in degree mode we shall use these familiar angles to enter a number x, press the sin button and then press inverse sin. The following table of results shows the outcome.

x	$y = \sin x$	$\sin^{-1} y$
$30°$	0.5	$30°$
$390°$	0.5	$-30°$
$360°$	0	$0°$
$180°$	0	$0°$
$90°$	1	$90°$
$-90°$	-1	$-90°$
$210°$	-0.5	$-30°$

This is just a sample of results of course. Trying some more soon provides convincing evidence that this process always gives a final result between $-90°$ (i.e., $-\frac{\pi}{2}$) and $90°$ (i.e., $\frac{\pi}{2}$).

Now if we restrict the sine function to the domain $\frac{\pi}{2} \leq x \leq \frac{\pi}{2}$ it is strictly increasing, and therefore 1-1, as Figure 1.21 shows.

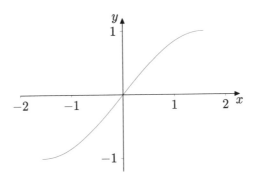

Figure 1.21 $y = \sin x$ for $\frac{\pi}{2} \leq x \leq \frac{\pi}{2}$

Readers should carry out similar exercises with cosine and tangent. It will be found that with inverse cosine the calculator will always return a value between 0 and π, and for the inverse tangent between $-\frac{\pi}{2}$ and $\frac{\pi}{2}$. Drawing graphs will show that in these intervals cosine and tangent respectively are 1-1 functions.

In each case the domain has been chosen so that the function when restricted to that domain is increasing (in the case of sine and tangent) or decreasing (in

the case of cosine), thereby ensuring that the function is 1-1. We shall therefore adopt the convention that the inverse function is always associated with
these intervals unless we specify a domain. We have seen that this convention
is programmed into calculators, and it is also the convention which MAPLE
adopts.

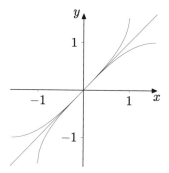

Figure 1.22   sin and its inverse

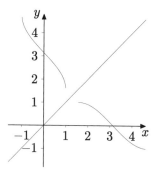

Figure 1.23   Another part of sin and its inverse

To summarise, the function for which we use the notation $\sin^{-1} x$ is the
inverse of the function specified by $f(x) = \sin x$, $-\frac{\pi}{2} \le x \le \frac{\pi}{2}$. An alternative
notation for the inverse is $\arcsin x$, and this is what MAPLE uses. Graphs of
$f(x)$ and its inverse are shown in Figure 1.22.

Now consider the function specified by $g(x) = \sin x$; $\frac{\pi}{2} \le x \le \frac{3\pi}{2}$. This is
a different function from $f(x)$. It is a decreasing function, and so it too has an
inverse, shown in Figure 1.23. In terms of $\arcsin x$ the formula for the inverse
in this case is $\pi + \arcsin(-x)$. Can you explain why?

Example 1.39

Find a formula for $\tan\left(\sin^{-1} x\right)$.

This is an example of a number of similar relationships involving trigonometric functions and their inverses. It can be derived algebraically using trigonometric identities as follows.

$$\tan\left(\sin^{-1} x\right) = \frac{\sin\left(\sin^{-1} x\right)}{\cos\left(\sin^{-1} x\right)} = \frac{\sin\left(\sin^{-1} x\right)}{\sqrt{1 - \sin^2\left(\sin^{-1} x\right)}} = \frac{x}{\sqrt{1 - x^2}}.$$

A more transparent method is to use a right-angled triangle. We therefore draw a right-angled triangle with an angle whose sin is x, as shown in Figure 1.24, where $\sin\theta = BC/CA = x$.

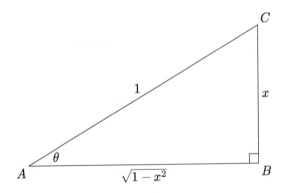

Figure 1.24   Diagram for Example 1.39

We then calculate the third side of the triangle (AB in the figure) using Pythagoras' Theorem. We can then read off any of the trigonometric ratios we need. So we have $\tan\left(\sin^{-1} x\right) = \dfrac{BC}{AB} = \dfrac{x}{\sqrt{1 - x^2}}$.

In both the algebraic and geometrical approach we have assumed that we should use the positive square root, and so this needs discussion. If $0 < x < 1$ then $0 < \sin^{-1} x < \frac{\pi}{2}$, and so $\tan\left(\sin^{-1} x\right) > 0$. If $-1 < x < 0$ then we have $-\frac{\pi}{2} < \sin^{-1} x < 0$, and so $\tan\left(\sin^{-1} x\right) < 0$. In both cases therefore the positive square root gives the correct sign for $\dfrac{x}{\sqrt{1 - x^2}}$. If $x = 0$ then the problem does not arise because $\tan\left(\sin^{-1} 0\right) = \dfrac{0}{\sqrt{1 - 0^2}} = 0$. Finally $x = 1$ and $x = -1$ do not belong to the domain of $\dfrac{x}{\sqrt{1 - x^2}}$ or $\tan\left(\sin^{-1} x\right)$.

1.7.3 Inverse Hyperbolic Functions

The situation here is much less complicated than for inverse trigonometric functions. In fact sinh and tanh are 1-1, as seen in Figures 1.14 and 1.15, and so have inverses with no restriction on their domains. However, cosh is not 1-1, but has at most two values of x in the domain corresponding to a given value of y in the range. Figure 1.13 shows this graphically.

Because the hyperbolic functions are defined in terms of exponentials, we might expect their inverses to be expressible in terms of the inverse of the exponential function, namely the logarithmic function, and that is indeed the case. We shall explore the slightly more difficult case of the inverse of cosh, leaving sinh and tanh to the reader.

Example 1.40

Show that $\cosh^{-1} x = \ln\left(x + \sqrt{x^2 - 1}\right)$.

The convention in this case is that $\cosh^{-1}$ is used to denote the inverse of cosh with domain restricted to the non-negative numbers, which gives the increasing portion of the graph of $\cosh x$, to the right of the y-axis. This convention is programmed into calculators and MAPLE.

So let $y = \cosh x = \dfrac{e^x + e^{-x}}{2}$. Multiplying both sides by $2e^x$ and collecting terms gives $(e^x)^2 - 2ye^x + 1 = 0$. This is a quadratic equation for e^x with solutions $e^x = y \pm \sqrt{y^2 - 1}$. Therefore $x = \ln\left(y \pm \sqrt{y^2 - 1}\right)$.

The problem we have to sort out now is which of the two solutions corresponds to the inverse we require. Since $y = \cosh x$ we must have $y \geq 1$. We then have $y - \sqrt{y^2 - 1} \leq 1$ and $y + \sqrt{y^2 - 1} \geq 1$, so that $\ln\left(y - \sqrt{y^2 - 1}\right) \leq 0$ and $\ln\left(y + \sqrt{y^2 - 1}\right) \geq 0$. It is the latter we require because we have restricted x to the non-negative numbers so as to specify a 1-1 function. We have therefore shown the following.

The inverse of $f(x) = \cosh x$, $x \geq 0$ is given by $f^{-1}(x) = \ln\left(x + \sqrt{x^2 - 1}\right)$.
The inverse of $g(x) = \cosh x$, $x \leq 0$ is given by $g^{-1}(x) = \ln\left(x - \sqrt{x^2 - 1}\right)$.

The corresponding formulae for the other inverse hyperbolic functions, left as an exercise for the reader, are

$$\sinh^{-1} x = \ln\left(x + \sqrt{x^2 + 1}\right); \qquad \tanh^{-1} x = \frac{1}{2}\ln\left(\frac{1 + x}{1 - x}\right).$$

1.8 Piecewise Functions

In Section 1.6.3 we gave the definition of the modulus function. That definition contained the single formula $|x|$ and also specified the function "in pieces", according to whether x is positive or negative. Figure 1.20 illustrates another function defined in pieces. This (optional) section discusses this further, in the form of an investigation for the reader.

In the eighteenth century mathematicians were firmly wedded to the idea that a function must be given by a single formula, and thought that if it was defined for x in some interval then its values outside that interval were completely determined. During the nineteenth century the idea of function was considerably broadened, as the following quotation from the German mathematician Dirichlet (1805–1859) makes clear.

> It is not necessary that y be subject to the same rule as regards x throughout the interval; indeed one need not even be able to express the relationship through mathematical operations ... different parts (*of the function*) may be given by different laws, or ... entirely lawlessly. If a function is specified only for part of an interval, the manner of its continuation for the rest of the interval is entirely arbitrary.

Draw the graphs of $y = |x|$ and $y = x$ on the same axes. Can you explain from the graphs why $|x| + x = 0$ if x is negative? What is the result of adding $|x|$ and x if x is positive? As a result of this you should be able to see that

$$f(x) = |x| + x = \begin{cases} 2x & \text{if } x \geq 0, \\ 0 & \text{if } x < 0. \end{cases}$$

Like $|x|$ itself, we have here another example of a function defined in two pieces, which can also be represented as a single formula.

What happens if you subtract $|x|$ from x? (Let $g(x) = x - |x|$.)

See if you can construct a single formula involving $|x|$ to represent the following function.

$$h(x) = \begin{cases} 2x & \text{if } x \geq 0, \\ 3x & \text{if } x < 0. \end{cases}$$

A function such as this can be plotted using MAPLE, so you can check your own formula. We use the following command.

```
plot(piecewise(x>=0,2x,3x),x=-2..2);
```

The logical syntax of this command can be paraphrased as follows.

"Plot a function in pieces. If $x \geq 0$ plot $2x$, otherwise plot $3x$."

There are various adaptations of this command, which can be seen on the `?piecewise;` help page in MAPLE.

What graph will you get if you multiply $f(x)$ above by x? And what will you get if you multiply $g(x)$ above by $-x$? Now try to find a single formula to represent the following function.

$$k(x) = \begin{cases} x^2 & \text{if } x \geq 0, \\ x^3 & \text{if } x < 0. \end{cases}$$

As before you can check your answer by plotting your formula in MAPLE and verifying that the same graph is obtained using

`plot(piecewise(x>=0,x^2,x^3),x=-2..2);`

This is the beginning of an investigation you can pursue for yourself, perhaps by exploring question such as

1. What other pairs of functions can be joined together with a single formula?

2. Can we join pairs of functions together at some other point (e.g., at $x = 1$ instead of at $x = 0$)?

3. Can we build up single formulae for functions consisting of three (or more) pieces?

It is a worthwhile exercise to write up the results of your investigation in a form suitable for a student mathematics magazine.

EXERCISES

1.1. Find the maximal domain for the function defined by each of the following expressions.

(a) $\tan 2x$;   (b) $x^5 - 3x^2 + 2x - 7$;   (c) $\ln(1 - x)$;

(d) $\ln\left(1 - x^2\right)$;   (e) $\dfrac{1}{x^2 - 5x + 6}$;   (f) $|\ln x|$;

(g) $\sqrt{x^2 - 3x - 4}$;   (h) $\dfrac{x}{1 - \sqrt{1 - x}}$;   (i) $\dfrac{1}{\ln\left(x^2 - 4\right)}$;

(j) $\dfrac{1}{\sin 2x}$;   (k) $\dfrac{1}{x\cos x}$;   (l) $\dfrac{1}{e^x - 1}$.

1.2. Write down the domains and ranges and sketch the graphs for

$$y = \sqrt{x}; \quad y = \sqrt{-x}; \quad y = -\sqrt{x}; \quad y = -\sqrt{-x}.$$

1.3. Find the maximal domain and range of $f(x) = \dfrac{2x + 3}{x - 5}$.

1.4. Plot the graphs of the functions defined in Exercise 1.1 using a graphical calculator, or MAPLE, or some other package. Use the graphs to write down what you think the range of the function is in each case.

1.5. Classify the functions defined by each of the following expressions as even, odd or neither.

(a) $5x^4 - 3x^2$;   (b) $2x^4 - x^3 + 1$;   (c) $\sin(x^3)$;

(d) $\sin^2 x^3$;   (e) $e^{|x|}$;   (f) $\ln|x|$;

(g) $\tan(\sin x)$;   (h) $e^{\sin x}$;   (i) $\sin(\ln x)$.

1.6. Prove that if f and g are odd functions then the product fg is an even function. Prove also that the product of two even functions is even, and that the product of an even function and an odd function is an odd function.

1.7. Let $f(x) = \dfrac{3}{x}$; $g(x) = \dfrac{2-x}{2+x}$. Construct and simplify the composite functions $f \circ f$, $f \circ g$, $g \circ f$, $g \circ g$. Write down the domain for each.

1.8. Sketch the graph of $y = e^{\sin x}$, using the method of Example 1.10, giving a clear explanation of your reasoning.

1.9. Given that $f(x) = e^x$ and $f \circ g(x) = 3x - 4$, find a formula for $g(x)$.

1.10. Find the function $f(x)$ satisfying $f \circ g(x) = \dfrac{1}{x^2}$ and $g(x) = 2x + 1$.

1.11. Carry out the following polynomial divisions.

(a) $\dfrac{x^4 + 2x^2 + 1}{x^4 - 2x^2 - 1}$;

(b) $\dfrac{x^3 + 1}{x + 2}$;

(c) $\dfrac{x^4 - x^3 + 2x^2 - 7}{2x + 5}$;

(d) $\dfrac{x^5 - 4x^3 + x - 1}{x^2 + x - 3}$;

(e) $\dfrac{x^6 - 1}{x^2 - x - 2}$;

(f) $\dfrac{x^5 - x^4 + 2x^3 + x}{x^3 + 2x^2 - 3}$.

1.12. Factorise the following polynomials

(a) $x^3 + x^2 - 4x - 4$;   (b) $y^3 + y^2 - y - 1$;

(c) $z^3 + 2z^2 - 23z - 60$;   (d) $c^5 + c^4 - 2c^3 - 2c^2 + c + 1$;

(e) $t^6 - 1$;   (f) $u^4 + u^3 - 25u^2 - 37u + 60$.

1.13. Show that $x = -2$ is a root of $g(x) = x^5 + 9x^4 + 32x^3 + 56x^2 + 48x + 16$. Without factorising $g(x)$ find the multiplicity of this root.

1.14. Solve the following equations and inequalities.

(a) $|2x - 5| = 4$;　(b) $|x^2 - 2x - 5| = 1$;　(c) $|x^3 - 1| = 7$;

(d) $|2x + 4| < 3$;　(e) $|2x^2 - 5x - 4| \leq 3$;　(f) $|x^2 + 5x| \geq 2$.

1.15. Sketch the graphs of the functions defined by the following expressions.

(a) $|x^2 - 5x + 6|$;　(b) $|\sin 2x|$;　(c) $\left| \dfrac{x - 2}{x + 1} \right|$.

[Hint: Sketch the graphs without the modulus first.]

1.16. Find an identity for $\cos 5x$ in terms of $\cos x$.

1.17. Find an addition formula for $\operatorname{cosec}(x + y)$ in terms of sec and cosec.

1.18. Use the laws of indices for the exponential function to deduce that

$$\ln\left(\frac{p}{q}\right) = \ln p - \ln q.$$

1.19. Simplify the following expressions as far as possible.

(a) $\left(\frac{1}{3}\right)^x 9^{x/2}$;　　　　(b) $\log_9\left(\frac{1}{27}\right)$;

(c) $\ln(1 + \cos x) + \ln(1 - \cos x) - 2\ln(\sin x)$.

1.20. Find an identity for $\sinh 3x$ in terms of $\sinh x$.

1.21. Starting from the graphs of sinh, cosh and tanh, sketch their reciprocals: cosech, sech and coth.

1.22. Simplify the following expressions as far as possible.

(a) $\cosh(\ln x)$;　　　　(b) $\coth(\ln x)$;

(c) $\dfrac{\cosh(\ln x) - \sinh(\ln x)}{\cosh(\ln x) + \sinh(\ln x)}$.

1.23. For each of the following functions $f(x)$, write down the domain, prove that it is 1-1, find a formula for the inverse $f^{-1}(x)$, and write down the domain of the inverse. For each function sketch its graph and the graph of its inverse on the same axes.

$$\text{(a)}\ \ f(x) = \frac{2x}{3x-1}; \quad \text{(b)}\ \ f(x) = \frac{2x}{\sqrt{x^2+3}}.$$

1.24. For each of the following functions, whose domain is specified, write down whether you think it is 1-1, by considering the formula. Plot the graph of each function to confirm your conclusions (or otherwise).

(a) $\tan x$, $0 \le x \le \frac{\pi}{4}$;   (b) $e^{|x|}$, $-1 \le x \le 1$;

(c) $\cosh 2x$, $x \ge 1$;   (d) $x^2 + 6x + 9$, $x \ge 0$;

(e) $x^2 - 6x + 9$, $x \ge 0$;   (f) $1/(x-1)$, $x \ne 1$;

(g) $1/(x-1)^2$, $x > 1$;   (h) $\exp(x)\ln(x)$, $x > 0$.

1.25. Prove that if $f(x)$ and $g(x)$ are increasing functions with domain D, then $f(x) + g(x)$ is an increasing function. Give examples to show that $f(x) - g(x)$, $f(x)g(x)$ and $f(x)/g(x)$ are not necessarily increasing functions.

1.26. Find a formula for the inverse of the following function.

$$f(x) = 5 - 12x - 2x^2, \quad x \ge -3.$$

Sketch the graph of $f(x)$ and the graph of its inverse on the same axes.

1.27. Simplify the following expressions as far as possible.

(a) $\cos\left(\sin^{-1} x\right)$;   (b) $\sin\left(\tan^{-1} x\right)$;   (c) $\tan\left(\sec^{-1} x\right)$.

1.28. Find numerical examples to show that in general $\tan^{-1} x \ne \dfrac{\sin^{-1} x}{\cos^{-1} x}$.

1.29. Starting from the graphs of sinh, cosh and tanh, sketch the graphs of their inverses by reflection in the line $y = x$.

1.30. Prove the formulae given at the end of Section 1.7.3 for $\sinh^{-1} x$ and $\tanh^{-1} x$.

1.31. In electrical circuit theory there is a function called the Heaviside function, named after the physicist Oliver Heaviside (1850–1925), defined as follows.

$$H(t) = \begin{cases} 1 & \text{if } t \geq 0, \\ 0 & \text{if } t < 0. \end{cases}$$

It represents the input to a circuit being switched on at time $t = 0$. So here again we have a function defined in two pieces for which a single symbol is used.

What does $H(-t)$ represent? What does $H(t-1)$ represent? What is the graph of $H(t) - H(t-1)$?

You can use this as a starting point for another investigation into piecewise functions having single formulae. These functions are in fact used in connection with Laplace Transforms for solving differential equations connected with electrical circuits.

2

Limits of Functions

2.1 What are Limits?

The ideas of limits are used in both theory and applications of calculus. You may have used limiting ideas graphically in discussing such things as asymptotes for the graph of a function. Developing a precise definition of limit took mathematicians around 200 years, mainly during the 18th and 19th centuries, in parallel with research on continuous functions. Much of this work arose from investigations on problems involving vibrations and related physical phenomena. It was found that the formulae developed for representing such phenomena gave rise to graphs with sudden jumps or discontinuities. To analyse these mathematically and to prove useful results it was necessary to develop a precise definition of continuous and discontinuous functions. This led to the development of the subject we now study as Real Analysis, discussed in the book with that title by John Howie in this series.

This chapter is not an exhaustive treatment of the theory of limits. We begin, in the first three sections, by considering a number of examples which illustrate graphically the ideas of limits, including discussion of asymptotes. Then we list, without proof, rules which help to calculate limits algebraically. Subsequent sections adopt a more systematic approach to such calculations, based on various kinds of algebraic methods. We aim to cover a sufficient range of situations and examples to facilitate an understanding of limits from the graphical and algebraic points of view needed in studying calculus.

From an informal point of view we can think of a continuous function as one whose curve can be drawn without taking one's pencil off the paper. That is an adequate description from a visual point of view, but less helpful in discussing functions given by algebraic formulae. Related to this however is the idea that if we have a function $f(x)$ then it is continuous if a small change in x results in a correspondingly small change in the value of $f(x)$. A mathematical analysis of the intuitive idea of small changes leads to the present-day definitions used in Real Analysis. If we put these ideas together we can think of what happens to the value of $f(x)$ as x gets closer and closer to some fixed number a, and we look for some definite number l which the values of $f(x)$ approach. When x is very close to a this corresponds to the small changes in values discussed above; using the word "approach" suggests motion along a graph, where $f(x)$ changes gradually without any sudden breaks as x also changes gradually.

So what sort of formulae correspond to continuous functions? You will have had sufficient experience of graphs to realise that most straightforward formulae do lead to continuous graphs unless there are values of x for which some term in a denominator is zero. So polynomials, exponential functions, sine and cosine, square root are examples of functions with continuous curves. Combinations of continuous functions, using operations such as addition, multiplication and composition, are also continuous; this is one of the results proved in Howie, Chapter 3.

Functions like the tangent function in trigonometry, and reciprocal functions, are examples which have discontinuities. Using the definition of the tangent function as $\tan x = \dfrac{\sin x}{\cos x}$, we would suspect that there is a discontinuity for all values of x for which $\cos x = 0$. Indeed such points do not belong to the domain. The graphs of tan and cot in Figure 1.12 show such discontinuities clearly.

Example 2.1

A further example of discontinuity is illustrated in Figure 2.1. The denominator of $\dfrac{1}{x^2 - 1}$ is zero when $x = \pm 1$, and this indicates a discontinuity at each of these points, as Figure 2.1 shows. The numbers $x = \pm 1$ are not in the domain of the function. More examples of graphs exhibiting this kind of behaviour can be seen in Figures 1.2 and 1.7.

When we have a continuous function, as x approaches a the value of $f(x)$ approaches the value of the function at $x = a$, namely $f(a)$. If $f(x)$ had a limiting value different from $f(a)$ as x approached a, that would indicate a discontinuous jump in the graph.

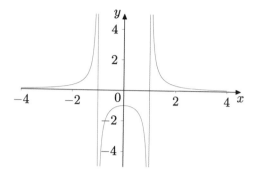

Figure 2.1   Graph of $\frac{1}{x^2-1}$

Example 2.2

In many situations where we discuss limits, we are looking at the behaviour of a function near to a number which does not belong to the domain of the function, i.e., for which the function is undefined. For example the function $f(x) = \dfrac{\sin x}{x}$ is not defined at $x = 0$. However if we plot its graph it appears to suggest that the value of $f(x)$ approaches 1 as x becomes ever closer to zero. This is discussed further in Example 2.12. In fact MAPLE plots the graph in such a way as to make it appear continuous even at $x = 0$, as Figure 2.2 shows.

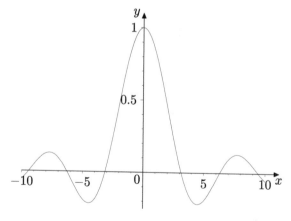

Figure 2.2   Graph of $\frac{\sin x}{x}$

In cases such as this where a function $f(x)$ tends to a limit, l say, as x tends to a number a which is not in the domain of f, we can extend the definition of the function to include this point. With the example above, we would write

$$f(x) = \begin{cases} \frac{\sin x}{x} & \text{for } x \neq 0, \\ 1 & \text{for } x = 0. \end{cases}$$

We say that the definition of $f(x)$ has been **extended by continuity**.

Example 2.3

This example looks at the behaviour of the function defined by $f(x) = \sin\left(\frac{1}{x}\right)$,
where the point $x = 0$ does not belong to the domain. If we plot the graph,
as in Figure 2.3, we see that the behaviour near to $x = 0$ appears erratic.
This is an example where the limitations of any plotting device are apparent.
The program calculates the value of the function at a finite number of points
and joins them up. With most functions the result appears to be a smooth
curve, but not in this case. In fact the graph oscillates between ± 1 infinitely
many times as x approaches zero. The peaks of all the oscillations should be
on the line $y = 1$ rather than the somewhat variable positions indicated on the
MAPLE plot. We need to undertake some calculations to establish this.

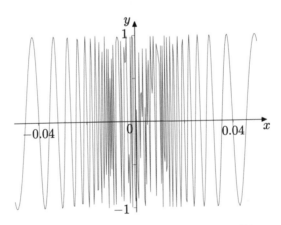

Figure 2.3   MAPLE plot of $\sin\left(\frac{1}{x}\right)$

Firstly, because $-1 \leq \sin t \leq 1$, we can be certain that $-1 \leq \sin\left(\frac{1}{x}\right) \leq 1$.

Next we shall see where the graph crosses the x-axis. Now $\sin t = 0$ when
$t = n\pi$, for all integers n. Therefore $\sin\left(\frac{1}{x}\right) = 0$ when $\frac{1}{x} = n\pi$, i.e., $x = \frac{1}{n\pi}$
(provided $n \neq 0$). This is a sequence of numbers which tends towards zero as
n increases.

Now $\sin t = 1$ when $t = \dfrac{(4n+1)\pi}{2}$, for all integers n. Therefore $\sin\left(\dfrac{1}{x}\right) = 1$

when $\dfrac{1}{x} = \dfrac{(4n+1)\pi}{2}$, i.e. $x = \dfrac{2}{(4n+1)\pi}$. This is also a sequence of numbers which tends towards zero as n increases.

Finally $\sin t = -1$ when $t = \dfrac{(4n-1)\pi}{2}$, for all integers n. Hence we have

$\sin\left(\dfrac{1}{x}\right) = -1$ when $\dfrac{1}{x} = \dfrac{(4n-1)\pi}{2}$, i.e. $x = \dfrac{2}{(4n-1)\pi}$. Again this is a

sequence of numbers which tends towards zero as n increases.

This proves that the graph does indeed oscillate between 1 and -1 infinitely many times as x approaches zero from either direction, since n can be positive or negative. So however small an interval containing zero we consider there are values of x inside that interval where $f(x) = 0$, where $f(x) = 1$ and where $f(x) = -1$. So $f(x)$ does not tend to a limiting value as x tends to zero.

Figure 2.3 is rather erratic, because as we have shown there are infinitely many oscillations in any interval including the origin. We can get a clearer picture of the oscillations if we use a domain which does not include $x = 0$, as in Figure 2.4.

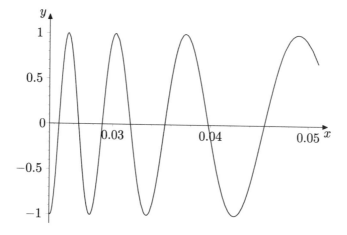

Figure 2.4   MAPLE plot of $\sin\left(\dfrac{1}{x}\right)$

Definition 2.4

We summarise the discussion and examples above in this informal definition of a limit, and introduce the two notations most commonly used for limits.

The function $f(x)$ is said to have the **limit** l as x approaches (or tends to) a if the values of $f(x)$ can be made as close as we like to l by taking x sufficiently close to a. We use the two notations

$\lim\limits_{x \to a} f(x) = l$, read as "the limit as x tends to a of $f(x)$ is equal to l",

$f(x) \to l$ as $x \to a$, read as "$f(x)$ tends to l as x tends to a".

It is assumed here that the domain of $f(x)$ includes an interval containing a, but not necessarily a itself.

The function $f(x)$ is said to be **continuous** at $x = a$ if $\lim\limits_{x \to a} f(x) = f(a)$, or, in the other notation, if $f(x) \to f(a)$ as $x \to a$.

The much more precise definitions employed in Real Analysis, for example in Howie, Chapter 3, analyse in terms of sets of real numbers the phrases "as close as we like to" and "taking x sufficiently close to" used in the informal definition above.

2.2 One-sided Limits

Sometimes we encounter a situation where the behaviour of a function near to a differs, depending on whether x approaches a from below or above. The language and notation to express these ideas in general is as follows.
The limit of $f(x)$ as x tends to a from below (or from the left) is equal to l :

$\lim\limits_{x \to a^-} f(x) = l.$

The limit of $f(x)$ as x tends to a from above (or from the right) is equal to m :

$\lim\limits_{x \to a^+} f(x) = m.$

Example 2.5

A simple example is the so-called "floor" function, whose value at x is the greatest integer less than or equal to x. Part of its graph is shown in Figure 2.5. If we look for example at what happens near to $x = 2$, MAPLE indicates, with a small mark on the left-hand end of the horizontal line segment, that the value of the function at 2 itself is 2. Just to the right the greatest integer less than or equal to 2.01 for example is 2. Just to the left, the greatest integer less than or equal to 1.99 is 1. We have used solid and open dots to indicate this, another common convention.

So using the notation introduced above we can write

$$\lim\limits_{x \to 2^-} \text{floor}(x) = 1, \quad \lim\limits_{x \to 2^+} \text{floor}(x) = 2.$$

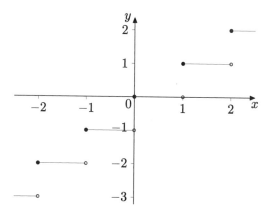

Figure 2.5   Graph of the floor function

At each integer value of x the floor function has a discontinuity, called a "jump discontinuity" because the function jumps between two values as x moves across the point of discontinuity. This contrasts with the discontinuity of $f(x) = \sin\left(\dfrac{1}{x}\right)$, discussed in Example 2.3 above. In that case neither the left-hand limit nor the right-hand limit exists.

2.3 Infinite Limits and Limits at Infinity

In considering graphs we often want to decide what happens to the graph when x or y increases or decreases without bounds. When we want to discuss the behaviour of a function $f(x)$ as x increases without an upper bound we use the phrase "x tends to infinity", symbolised by $x \to \infty$. When x decreases without bound, becoming very large and negative, we use the phrase "x tends to minus infinity", denoted by $x \to -\infty$. It is important to emphasise that the symbol ∞ does not represent a real number.

The following examples illustrate the use of this language and notation.

Example 2.6

In Figure 2.6 we observe that the graph of the function $x^2 - 1$ appears to increase without bound as x tends to infinity (and to minus infinity). Using the notation introduced above for limits, we write

$$x^2 - 1 \to \infty \text{ as } x \to \infty, \qquad x^2 - 1 \to \infty \text{ as } x \to -\infty.$$

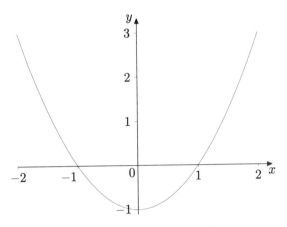

Figure 2.6   Graph of $x^2 - 1$

If we consider the graph of $y = x^3 + 1$ shown in Figure 2.7 we describe the limiting behaviour using the notation

$$x^3 + 1 \to \infty \ \text{ as } \ x \to \infty, \qquad x^3 + 1 \to -\infty \ \text{ as } \ x \to -\infty.$$

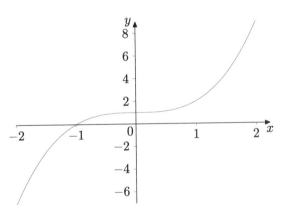

Figure 2.7   Graph of $x^3 + 1$

Example 2.7

In the example in Figure 2.8 the function has finite limits as x tends to infinity and as x tends to minus infinity. The two notations in this case are

$$\frac{1 + 3e^x}{1 + e^x} \to 3 \ \text{ as } \ x \to \infty, \qquad \frac{1 + 3e^x}{1 + e^x} \to 1 \ \text{ as } \ x \to -\infty.$$

$$\lim_{x \to \infty} \frac{1 + 3e^x}{1 + e^x} = 3, \qquad \lim_{x \to -\infty} \frac{1 + 3e^x}{1 + e^x} = 1.$$

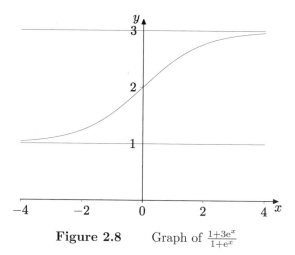

Figure 2.8   Graph of $\frac{1+3e^x}{1+e^x}$

Finally we consider examples where the function is unbounded in the neighbourhood of some point on the x-axis.

Example 2.8

Consider the graph shown in Figure 2.9.

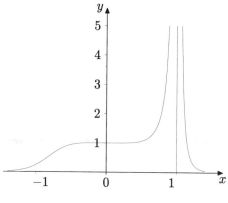

Figure 2.9   Graph of $\frac{1}{(x^5-1)^2}$

We use the notation

$$\frac{1}{(x^5 - 1)^2} \to \infty \text{ as } x \to 1.$$

In this case the limiting behaviour is the same to the left and to the right of $x = 1$.

If we look at the example shown in Figure 2.1 we see that the behaviour is not the same on each side of $x = 1$, or of $x = -1$. We have different one-sided limiting behaviour, symbolised by

$$\frac{1}{x^2 - 1} \to \infty \text{ as } x \to 1^+, \qquad \frac{1}{x^2 - 1} \to -\infty \text{ as } x \to 1^-,$$

$$\frac{1}{x^2 - 1} \to \infty \text{ as } x \to -1^-, \qquad \frac{1}{x^2 - 1} \to -\infty \text{ as } x \to -1^+.$$

We used the idea of an asymptote graphically in a number of examples in Chapter 1. We can now give a description in terms of limits. In Figure 2.8 the graph approaches the horizontal line $y = 3$ as x tends to infinity, and it approaches $y = 1$ as x tends to minus infinity. In Figure 2.1 the graph approaches the vertical line $x = 1$ as x tends to 1 from above and below, with similar behaviour near $x = -1$. This gives rise to the following definition.

Definition 2.9

If $f(x) \to m$ as $x \to \infty$ (or as $x \to -\infty$) then we say that the line $y = m$ is a **horizontal asymptote**. (This includes the possibility $m = 0$, in which case the x-axis is a horizontal asymptote.)

If $f(x) \to \pm\infty$ as $x \to a^+$ or as $x \to a^-$ then we say that the line $x = a$ is a **vertical asymptote**. (This includes the possibility $a = 0$, in which case the y-axis is a vertical asymptote.)

We can see further examples of horizontal asymptotes in Figures 1.2, 1.7, 1.15, 2.1, 2.9. In Figure 2.2, the x-axis is a horizontal asymptote. This is clearer if the graph is plotted for a larger domain, and if the graphs of $y = \pm\frac{1}{x}$ are added. The MAPLE command to plot this is

```
plot([sin(x)/x,1/x,-1/x],x=-50..50,y=-0.25..1,color=black);
```

Vertical asymptotes are shown in Figures 1.2, 1.3, 1.5, 1.7, 1.10, 1.11, 1.12, 2.1.

2.4 Algebraic Rules for Limits

Example 2.10

Find $\lim\limits_{x \to 1} \left(x^2 + \dfrac{x^3 + 1}{x^2 + 1} + 2x \sin\left(\pi\sqrt{3x^2 + 1}\right) \right)$.

Since the denominator $x^2 + 1$ is never zero, our knowledge of expressions like this and their graphs should tell us that whatever its precise shape, the graph will be continuous, so the limit can just be found by substituting $x = 1$ giving

$$1^2 + \frac{1^3 + 1}{1^2 + 1} + 2 \sin\left(\pi\sqrt{3 + 1}\right) = 1 + 1 + 2 \sin 2\pi = 2.$$

We have implicitly used the following rules for limits

If $f(x) \to l$ as $x \to a$ and $g(x) \to m$ as $x \to a$ then

$f(x) + g(x) \to l + m$ as $x \to a$ (addition rule);

$f(x) - g(x) \to l - m$ as $x \to a$ (subtraction rule);

$f(x)g(x) \to lm$ as $x \to a$ (multiplication rule);

$\dfrac{f(x)}{g(x)} \to \dfrac{l}{m}$ as $x \to a$ (provided $m \neq 0$) (division rule).

If $f(t) \to l$ as $t \to a$ and $g(x) \to a$ as $x \to b$

then $f(g(x)) \to l$ as $x \to b$ (composition rule).

Each of the above rules has an analogue for one-sided limits, for example

If $f(x) \to l$ as $x \to a^+$ and $g(x) \to m$ as $x \to a^+$

then $f(x) + g(x) \to l + m$ as $x \to a^+$.

There is a similar set of rules for limits at infinity (and at minus infinity).

If $f(x) \to l$ as $x \to \infty$ and $g(x) \to m$ as $x \to \infty$ then

$f(x) + g(x) \to l + m$ as $x \to \infty$ (addition rule);

$f(x) - g(x) \to l - m$ as $x \to \infty$ (subtraction rule);

$f(x)g(x) \to lm$ as $x \to \infty$ (multiplication rule);

$\dfrac{f(x)}{g(x)} \to \dfrac{l}{m}$ as $x \to \infty$ (provided $m \neq 0$) (division rule).

If $f(t) \to l$ as $t \to a$ and $g(x) \to a$ as $x \to \infty$

then $f(g(x)) \to l$ as $x \to \infty$ (composition rule).

There are similar rules involving infinite limits for addition and multiplication, namely

$$\text{If} \quad f(x) \to \infty \text{ as } x \to a \quad \text{and} \quad g(x) \to \infty \text{ as } x \to a \quad \text{then}$$

$$f(x) + g(x) \to \infty \text{ as } x \to a \quad \text{(addition rule)};$$

$$f(x)g(x) \to \infty \text{ as } x \to a \quad \text{(multiplication rule)}.$$

However there are no analogous rules for infinite limits involving subtraction or division. A variety of outcomes is possible, depending on the functions involved, and this is where considerable care has to be taken. Examples 2.11, 2.21 and 2.22 demonstrate this.

Example 2.11

Let $f(x) = \dfrac{1}{x^2}$ and $g(x) = \dfrac{1}{x^4}$. We can see, by plotting graphs for example, that $f(x) \to \infty$ as $x \to 0$ and $g(x) \to \infty$ as $x \to 0$. Then $\dfrac{f(x)}{g(x)} = x^2 \to 0$ as $x \to 0$. However, $\dfrac{g(x)}{f(x)} = \dfrac{1}{x^2} \to \infty$ as $x \to 0$.

In both cases the numerator and denominator tend to infinity, but in one case the quotient tends to zero and in the other case the quotient tends to infinity.

2.5 Techniques for Finding Limits

Example 2.11 involves expressions for which we cannot find limits by a simple application of algebraic rules. Examples of such types of expression are:

$$\dfrac{f(x)}{g(x)} \text{ where } f(x) \to 0 \text{ and } g(x) \to 0 \qquad \text{``}\tfrac{0}{0}\text{''};$$

$$\dfrac{f(x)}{g(x)} \text{ where } f(x) \to \infty \text{ and } g(x) \to \infty \qquad \text{``}\tfrac{\infty}{\infty}\text{''};$$

$$f(x) \times g(x) \text{ where } f(x) \to 0 \text{ and } g(x) \to \infty \qquad \text{``}0 \times \infty\text{''};$$

$$f(x) - g(x) \text{ where } f(x) \to \infty \text{ and } g(x) \to \infty \qquad \text{``}\infty - \infty\text{''};$$

$$f(x)^{g(x)} \text{ where } f(x) \to 1 \text{ and } g(x) \to \infty \qquad \text{``}1^\infty\text{''}.$$

We can find examples of functions f and g for which such limits are zero, infinity, negative infinity, finite and non-zero, or even non-existent. Readers are asked to explore this in an exercise at the end of this chapter.

We shall see in Chapter 3 that the definition of the derivative $f'(x)$ is $\lim_{h \to 0} \dfrac{f(x+h) - f(x)}{h}$. Both numerator and denominator of the quotient tend to zero, so this limit is a "$\frac{0}{0}$" type. It is clear therefore that such types of limit are central to the development of calculus.

We now investigate some of these types, illustrating them through examples rather than general theory.

We shall look at four techniques: squeezing, algebraic manipulation, change of variable, and l'Hôpital's Rule.

2.5.1 Squeezing

Example 2.12

We shall give a geometrical proof of the standard result $\lim_{x \to 0} \dfrac{x}{\sin x} = 1$. The domain of the expression does not include $x = 0$. The limit is a "$\frac{0}{0}$" type. Now $\dfrac{x}{\sin x}$ is an even function, so its behaviour as x tends to zero from above will be the same as that from below. Since we are considering the limit as $x \to 0$ we can assume that $0 < x < \dfrac{\pi}{2}$, as in Figure 2.10.

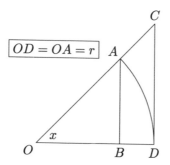

Figure 2.10   Diagram for $\lim_{x \to 0} \dfrac{x}{\sin x}$

In Figure 2.10 AD is a circular arc, and comparing areas we have

$$\Delta OAB < \text{ sector } OAD < \Delta OCD;$$
$$\tfrac{1}{2}(OB)(OA \sin x) < \tfrac{1}{2}OA^2 x < \tfrac{1}{2}(OD)(OC \sin x);$$
$$\tfrac{1}{2}r\, r \cos x \sin x < \tfrac{1}{2}r^2 x < \tfrac{1}{2}r\, \frac{r}{\cos x} \sin x.$$

Dividing by $\tfrac{1}{2}r^2$ and by $\sin x$, which we are assuming to be positive, gives

$$\cos x < \frac{x}{\sin x} < \frac{1}{\cos x}.$$

Now $\cos x \to 1$ as $x \to 0$, so $1 \le \lim_{x \to 0} \dfrac{x}{\sin x} \le 1$, i.e., $\lim_{x \to 0} \dfrac{x}{\sin x} = 1$.

This is an example of the "squeezing" technique. Figure 2.11 shows that the graph of $\dfrac{x}{\sin x}$ is "squeezed" between the graphs of $\cos x$ and $\dfrac{1}{\cos x}$ for $-1 \le x \le 1$.

Thus all three functions have the same limit as $x \to 0$.

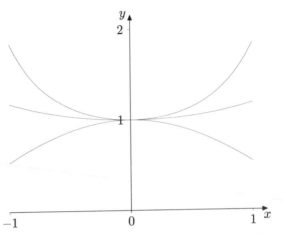

Figure 2.11   Squeezing graph for $\lim_{x \to 0} \dfrac{x}{\sin x}$

Example 2.13

Find $\lim_{x \to 0} x^2 \sin\left(\dfrac{1}{x}\right)$. This is an example where part of the expression, $\sin\left(\dfrac{1}{x}\right)$, has no limit at all, as we showed in Example 2.3. Nevertheless we can still use the squeezing technique. Remember that sin always lies between 1 and -1. We therefore have $-x^2 \le x^2 \sin\left(\dfrac{1}{x}\right) \le x^2$ for all $x \ne 0$. The outer expressions in this chain of inequalities both tend to zero as x tends to zero, and so the squeezing technique tells us that $\lim_{x \to 0} x^2 \sin\left(\dfrac{1}{x}\right) = 0$. We can see in Figure 2.12 that the graph is "squeezed" between the graphs of $y = \pm x^2$ as $x \to 0$.

This is yet another example where the limiting point ($x = 0$ in this case) does not belong to the domain, for we cannot substitute $x = 0$ in $\sin\left(\dfrac{1}{x}\right)$. However the fact that the limit is zero means that we can extend the definition of the function by continuity, as we discussed in Example 2.2.

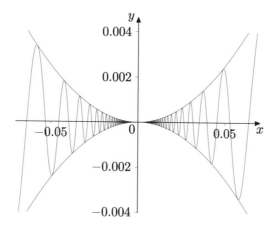

Figure 2.12   Graph of $x^2 \sin\left(\frac{1}{x}\right)$

2.5.2 Algebraic Manipulation

Example 2.14

Find $\lim\limits_{x \to 1} \dfrac{x^3 - 1}{x - 1}$.

We first factorise the numerator. We then observe that we can cancel the factor $(x - 1)$, which simplifies the expression. Finally we use algebraic rules of limits (as we shall throughout the following sections), calculating separately the limit of each term in the resulting expression. We therefore have

$$\lim_{x \to 1} \frac{x^3 - 1}{x - 1} = \lim_{x \to 1} \frac{(x - 1)(x^2 + x + 1)}{x - 1} = \lim_{x \to 1} (x^2 + x + 1) = 1 + 1 + 1 = 3.$$

The cancellation step needs further discussion. When we cancel we need to be sure that we are not dividing numerator and denominator by zero. In this case the factor $(x - 1)$ is zero only if $x = 1$. We are trying to find the limit as x tends to 1 and so $x = 1$ is not among the set of values under consideration. We are looking at values of x satisfying $1 - q < x < 1$ or $1 < x < 1 + q$ for some number $q > 0$, i.e., values of x in the neighbourhood of 1 but not including 1. In fact $x = 1$ does not belong to the domain of the function. Even if it did, the limit need not necessarily be the value of the function at that point, as Example 2.5 illustrates. This is an important aspect of limits, which occurs in many of the examples we shall consider.

The next few examples exemplify some of the results about rational functions which we considered in Section 1.6.2.

Example 2.15

Describe the behaviour of $\dfrac{x^3 - 2x + 3}{x^2 + 4}$ as $x \to \infty$ and as $x \to -\infty$.

The graph of this rational function was shown in Figure 1.6. The limit is an "$\frac{\infty}{\infty}$" type. The components of the expression which determine the behaviour when x is large are the greatest powers of x in the numerator and denominator. They "dominate" the rest of the terms. If we were to remove these other terms the resulting expression would simplify to x, and so we should expect the behaviour of the graph to be like that of $y = x$ as $x \to \pm\infty$. We demonstrate this algebraically by dividing numerator and denominator by x^2, the largest power of x in the denominator. This gives

$$\frac{x^3 - 2x + 3}{x^2 + 4} = \frac{x - \frac{2}{x} + \frac{3}{x^2}}{1 + \frac{4}{x^2}}.$$

The second and third terms in the numerator and the second term in the denominator tend to zero as $x \to \pm\infty$. The denominator therefore tends to 1. The numerator tends to ∞ as $x \to \infty$, and to $-\infty$ as $x \to -\infty$. We therefore conclude that

$$\frac{x^3 - 2x + 3}{x^2 + 4} \to \infty \ \text{ as } \ x \to \infty \ \text{ and } \ \frac{x^3 - 2x + 3}{x^2 + 4} \to -\infty \ \text{ as } \ x \to -\infty.$$

Example 2.16

Describe the behaviour of $f(x) = \dfrac{x^2 - 2}{x^3 - 2x^2 - x + 2}$ as $x \to \infty$ and as $x \to -\infty$.

The graph of this rational function was shown in Figure 1.7. The limit is again an "$\frac{\infty}{\infty}$" type. As in example 2.15 we divide numerator and denominator by the greatest power of x in the denominator, in this case x^3, giving

$$\frac{x^2 - 2}{x^3 - 2x^2 - x + 2} = \frac{\frac{1}{x} - \frac{2}{x^3}}{1 - \frac{2}{x} - \frac{1}{x^2} + \frac{2}{x^3}}.$$

The denominator tends to 1 as $x \to \pm\infty$, since all the terms except the first tend to zero. Both terms in the numerator tend to zero as $x \to \pm\infty$. So we have the numerator tending to zero and the denominator tending to 1. The quotient therefore tends to zero. So we have shown algebraically that

$$\lim_{x \to \pm\infty} \frac{x^2 - 2}{x^3 - 2x^2 - x + 2} = 0.$$

Therefore, as described in Definition 2.9, the x-axis is a horizontal asymptote, which Figure 1.7 illustrates.

Example 2.17

Determine the limiting behaviour of $f(x) = \dfrac{(x + \sqrt{2})(x - \sqrt{2})}{(x + 1)(x - 1)(x - 2)}$ at the values of x where the denominator is zero.

This is the function considered in the previous example, shown in Figure 1.7, with the numerator and denominator in their factorised form. We therefore have to consider the behaviour relative to $x = -1, 1, 2$. For these values the numerator tends to $-1, -1, 2$ respectively. The denominator tends to zero. It is a common mistake to conclude that if the denominator tends to zero and the numerator tends to a non-zero limit, then the fraction itself tends to infinity. This error is sometimes written as "$\frac{1}{0} = \infty$". This is WRONG. It is clear from Figure 1.7 that at each of $x = -1, 1, 2$ we have different one-sided limiting behaviour. To correct the mistake we have to take careful account of the signs of the various parts of the expression. We can do this either by plotting the graphs of numerator and denominator, or by means of a tabular approach, as follows.

		$-\sqrt{2}$		-1		1		$\sqrt{2}$		2	
$x + \sqrt{2}$	$-$		$+$		$+$		$+$		$+$		$+$
$x - \sqrt{2}$	$-$		$-$		$-$		$-$		$+$		$+$
$x + 1$	$-$		$-$		$+$		$+$		$+$		$+$
$x - 1$	$-$		$-$		$-$		$+$		$+$		$+$
$x - 2$	$-$		$-$		$-$		$-$		$-$		$+$
$f(x)$	$-$		$+$		$-$		$+$		$-$		$+$

At the top of the table we have indicated the numbers where either the numerator or denominator is zero. At each of these points one of the factors changes sign, and so the table indicates the sign of each factor in the intervals between these numbers. At the bottom of the table we have shown the sign of $f(x)$ itself, obtained from the set of signs above according to whether there is a even or an odd number of negative signs. The numbers $\pm\sqrt{2}$ are zeros of the numerator, so the graph crosses the x-axis at those points, and $f(x)$ changes sign as indicated, and as the graph confirms.

Now let us consider what happens near to $x = -1$. We can see from the table that if $x < -1$ then $f(x) > 0$, whereas if $x > -1$ then $f(x) < 0$. In both cases the numerator tends to -1 and the denominator tends to zero. We therefore conclude that

$$f(x) \to \infty \text{ as } x \to -1^-, \text{ and } f(x) \to -\infty \text{ as } x \to -1^+.$$

Using similar reasoning near to $x = 1$ and $x = 2$ we conclude that

$$f(x) \to -\infty \text{ as } x \to 1^-, \text{ and } f(x) \to \infty \text{ as } x \to 1^+.$$

$$f(x) \to -\infty \text{ as } x \to 2^-, \text{ and } f(x) \to \infty \text{ as } x \to 2^+.$$

The lines $x = -1, 1, 2$ are vertical asymptotes.

Example 2.18

Find $\lim\limits_{x \to \infty} \dfrac{3x^2 + 4x + 4}{4x^2 + 3x + 2}$.

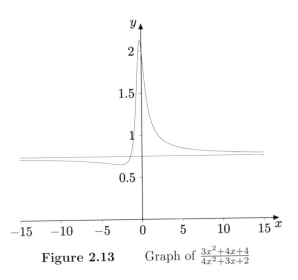

Figure 2.13   Graph of $\frac{3x^2+4x+4}{4x^2+3x+2}$

This is an "$\frac{\infty}{\infty}$" type. In this example highest power of x present is the same in numerator and denominator, namely x^2. We divide numerator and denominator by x^2, giving

$$\frac{3x^2 + 4x + 4}{4x^2 + 3x + 2} = \frac{3 + \frac{4}{x} + \frac{4}{x^2}}{4 + \frac{3}{x} + \frac{2}{x^2}} \to \frac{3 + 0 + 0}{4 + 0 + 0} = \frac{3}{4} \text{ as } x \to \infty.$$

So the line $x = \frac{3}{4}$ is a horizontal asymptote, shown in Figure 2.13. Solving the relevant quadratic equation shows that the denominator has no real zeros, and so there are no vertical asymptotes.

Example 2.19

Discuss the limiting behaviour of $\dfrac{1}{(x^5 - 1)^2}$.

The graph of this function can be seen in Figure 2.9. Firstly we observe that the denominator tends to infinity as $x \to \pm\infty$, and so the function itself tends to zero, i.e., the x-axis is a horizontal asymptote. Secondly, the denominator has its only real zero at $x = 1$. Because the denominator is a square, it is positive for all $x \neq 1$, and so the expression tends to infinity as x tends to 1 from both the left and the right. So we can write

$$\frac{1}{(x^5 - 1)^2} \to \infty \text{ as } x \to 1.$$

Example 2.20

Show that

$$\lim_{x \to -\infty} \frac{1 + 3e^x}{1 + e^x} = 1, \qquad \lim_{x \to \infty} \frac{1 + 3e^x}{1 + e^x} = 3.$$

The graph of this function can be seen in Figure 2.8. We first note that $e^x \to 0$ as $x \to -\infty$. Therefore using the rules for limits tells us that

$$\lim_{x \to -\infty} \frac{1 + 3e^x}{1 + e^x} \to \frac{1 + 0}{1 + 0} = 1.$$

To deal with the other limit, we use a similar division procedure as in the rational function examples above. This time we divide numerator and denominator by e^x. We use the fact that $e^{-x} \to 0$ as $x \to \infty$ and the rules of limits to give

$$\lim_{x \to \infty} \frac{1 + 3e^x}{1 + e^x} = \lim_{x \to \infty} \frac{e^{-x} + 3}{e^{-x} + 1} = \frac{0 + 3}{0 + 1} = 3.$$

In this example we have used our knowledge of the exponential function. In previous examples we needed to know the behaviour of powers of x as $x \to \infty$. Most of the examples in this section illustrate the principle of finding limits using knowledge of the limiting behaviour of a small number of basic functions, in this case powers and the exponential function.

Example 2.21

Find $\lim_{x \to 0} \dfrac{x}{\sqrt{1 - x} - \sqrt{1 + x}}$.

This is a "$\frac{0}{0}$" type, where the domain of the expression does not include zero. When dealing with algebraic expressions of this kind a common technique is that of rationalising the denominator, in this case giving

$$\frac{x}{\sqrt{1 - x} - \sqrt{1 + x}} = \frac{x\left(\sqrt{1 - x} + \sqrt{1 + x}\right)}{\left(\sqrt{1 - x} - \sqrt{1 + x}\right)\left(\sqrt{1 - x} + \sqrt{1 + x}\right)}$$

$$= \frac{x\left(\sqrt{1 - x} + \sqrt{1 + x}\right)}{(1 - x) - (1 + x)} = \frac{x\left(\sqrt{1 - x} + \sqrt{1 + x}\right)}{-2x}$$

$$= -\frac{\sqrt{1 - x} + \sqrt{1 + x}}{2} \to -1 \text{ as } x \to 0.$$

Example 2.22

Find the limit as $x \to \infty$ of each of the following expressions.

(a) $\left(\sqrt{x+1} - \sqrt{x}\right)$, (b) $\sqrt{x}\left(\sqrt{x+1} - \sqrt{x}\right)$, (c) $x\left(\sqrt{x+1} - \sqrt{x}\right)$.

(a) This is an example of an "$\infty - \infty$" type. In this case we rationalise the numerator, multiplying by the sum of square roots, and introducing this sum as a denominator, giving

$$\left(\sqrt{x+1} - \sqrt{x}\right) = \frac{\left(\sqrt{x+1} - \sqrt{x}\right)\left(\sqrt{x+1} + \sqrt{x}\right)}{\left(\sqrt{x+1} + \sqrt{x}\right)}$$

$$= \frac{(x+1-x)}{\left(\sqrt{x+1} + \sqrt{x}\right)} = \frac{1}{\left(\sqrt{x+1} + \sqrt{x}\right)}$$

$$\to 0 \ \text{as} \ x \to \infty,$$

because the denominator tends to infinity.

(b) The same rationalisation as in (a) gives

$$\sqrt{x}\left(\sqrt{x+1} - \sqrt{x}\right) = \frac{\sqrt{x}}{\left(\sqrt{x+1} + \sqrt{x}\right)}.$$

This expression is now an "$\frac{\infty}{\infty}$" type, and we divide the numerator and denominator by $\sqrt{x}$ to obtain

$$\sqrt{x}\left(\sqrt{x+1} - \sqrt{x}\right) = \frac{1}{\left(\sqrt{1 + \frac{1}{x}} + 1\right)} \to \frac{1}{2} \ \text{as} \ x \to \infty,$$

since the denominator tends to 2 as $x \to \infty$.

(c) Again we perform the same rationalisation, giving

$$x\left(\sqrt{x+1} - \sqrt{x}\right) = \frac{x}{\left(\sqrt{x+1} + \sqrt{x}\right)}.$$

Once more we have an "$\frac{\infty}{\infty}$" type, and dividing the numerator and denominator by $\sqrt{x}$ gives

$$x\left(\sqrt{x+1} - \sqrt{x}\right) = \frac{\sqrt{x}}{\left(\sqrt{1 + \frac{1}{x}} + 1\right)} \to \infty \ \text{as} \ x \to \infty,$$

since the numerator tends to ∞ and the denominator tends to 2.

This example illustrates the comment preceding Example 2.11, near the end of Section 2.4.

2.5.3 Change of Variable

We meet the technique of change of variable in many circumstances, for example in integration and differential equations. Here we can use it in limit problems. The general procedure is that we aim to use a substitution which will simplify the expression whose limit we are trying to find, ideally by reducing it to one whose limit we already know.

Example 2.23

Find $\lim\limits_{x \to \frac{\pi}{2}} \dfrac{\cos x}{\sin(\cos x)}$.

This is a "$\frac{0}{0}$" type, whose form should remind us of $\dfrac{x}{\sin x}$. This suggests that we substitute $t = \cos x$. Then $t \to 0$ as $x \to \frac{\pi}{2}$.

Therefore $\lim\limits_{x \to \frac{\pi}{2}} \dfrac{\cos x}{\sin(\cos x)} = \lim\limits_{t \to 0} \dfrac{t}{\sin t} = 1$, using the result of Example 2.12.

Notice that when we change the variable we have to pay attention to the limiting point, as in this case, where $x = \frac{\pi}{2}$ corresponds to $t = 0$. This relates directly to the composition rule for limits at infinity given in Section 2.4, which states that

if $f(t) \to l$ as $t \to a$ and $g(x) \to a$ as $x \to \infty$ then $f(g(x)) \to l$ as $x \to \infty$.

In this case we have applied the composition rule with

$$f(t) = \frac{t}{\sin t}, \quad l = 1, \ a = 0, \ g(x) = \cos x, \ b = \frac{\pi}{2}.$$

Example 2.24

Find $\lim\limits_{x \to 0} \dfrac{\tan^{-1} x}{x}$.

This is again a "$\frac{0}{0}$" type, and the composition rule can be used. A possible approach to simplifying the expression is to let $t = \tan^{-1} x$. So $x = \tan t$ and $t \to 0$ as $x \to 0$. Therefore

$$\lim\limits_{x \to 0} \frac{\tan^{-1} x}{x} = \lim\limits_{t \to 0} \frac{t}{\tan t} = \lim\limits_{t \to 0} \frac{t}{\sin t} \cdot \cos t = \lim\limits_{t \to 0} \frac{t}{\sin t} \cdot \lim\limits_{t \to 0} \cos t = 1,$$

using the result of Example 2.12 and the product rule for limits.

2.5.4 L'Hôpital's Rule

This rule is designed specifically to deal with "$\frac{0}{0}$" types and "$\frac{\infty}{\infty}$" types.

The rule involves differentiation, which is discussed in detail in Chapter 3. Readers who are unfamiliar with basic differentiation from school mathematics can omit this section until Chapter 3 has been studied. We shall give a proof in Section 6.2. At this stage we give the statement of the rule, including the conditions under which the rule is applicable, which must be understood. We shall look at some examples.

Statement of the Rule

Suppose that the functions f and g are both differentiable in an interval containing $x = a$. Suppose also that $f(x) \to 0$ as $x \to a$ and that $g(x) \to 0$ as $x \to a$. (**These are the conditions for the rule to be applicable and must always be checked.**)

If $\dfrac{f'(x)}{g'(x)} \to l$ as $x \to a$ then $\dfrac{f(x)}{g(x)} \to l$ as $x \to a$ (where l can be finite or infinite).

The rule applies with appropriate modifications for one-sided limits, for limits at infinity, and also when $f(x) \to \infty$ and $g(x) \to \infty$ as $x \to a$.

Example 2.25

Find $\displaystyle \lim_{x \to 1} \frac{\ln x}{x - 1}$.

We let $f(x) = \ln x$, $g(x) = x - 1$. We must first check that the conditions for the application of l'Hôpital's Rule are satisfied. Both f and g are differentiable near $x = 1$, and both tend to zero as x tends to 1. Now $f'(x) = \dfrac{1}{x}$, $g'(x) = 1$, so $\dfrac{f'(x)}{g'(x)} = \dfrac{1}{x} \to 1$ as $x \to 1$. Therefore by l'Hôpital's Rule $\dfrac{f(x)}{g(x)} \to 1$ as $x \to 1$.

We can make an interesting and important deduction from this example using a change of variable. If we let $x = 1 + \dfrac{a}{t}$ then $x \to 1$ as $t \to \infty$. So

$$
\begin{aligned}
1 &= \lim_{x \to 1} \frac{\ln x}{x - 1} = \lim_{t \to \infty} \frac{\ln\left(1 + \frac{a}{t}\right)}{\frac{a}{t}} \\
&= \lim_{t \to \infty} \frac{t \ln\left(1 + \frac{a}{t}\right)}{a} = \lim_{t \to \infty} \frac{\ln\left(1 + \frac{a}{t}\right)^t}{a}.
\end{aligned}
$$

Multiplying by a gives $\displaystyle \lim_{t \to \infty} \ln\left(1 + \frac{a}{t}\right)^t = a$, so $\displaystyle \lim_{t \to \infty} \left(1 + \frac{a}{t}\right)^t = e^a$.

There are many ways of defining the exponential function, and this limit is one of them. It is often encountered where t is restricted to integer values n, so that we can express $e^a = \lim_{n \to \infty} \left(1 + \dfrac{a}{n}\right)^n$ as the limit of a sequence of numbers.

Example 2.26

Find $\lim_{x \to 0^+} x \ln x$.

This is a "$0 \times \infty$" type, and so it is not immediately clear that we can apply l'Hôpital's Rule. However if we re-write the expression as

$$x \ln x = \frac{\ln x}{1/x}$$

then we can see it as an "$\dfrac{\infty}{\infty}$" type, to which the rule can be applied.

So we let $f(x) = \ln x$, $g(x) = \dfrac{1}{x}$. We then have $f(x) \to -\infty$ and $g(x) \to \infty$ as $x \to 0^+$. Calculating the derivatives gives

$$f'(x) = \frac{1}{x}, \quad g'(x) = -\frac{1}{x^2}, \quad \text{so} \quad \frac{f'(x)}{g'(x)} = \frac{1/x}{-1/x^2} = -x \to 0 \quad \text{as} \quad x \to 0^+.$$

Therefore by l'Hôpital's Rule, $\lim_{x \to 0^+} x \ln x = 0$.

Example 2.27

Find $\lim_{x \to 0} (\cos x)^{\frac{1}{x}}$.

This does not look like a "$\dfrac{0}{0}$" type. In fact as $x \to 0^+$ it is a 1^∞ type, and as $x \to 0^-$ it is a $1^{-\infty}$ type. Algebraic intuition does not help much here. On the one hand we might think that when $x = 0$, $\cos x = 1$, and 1 raised to any power is always 1, so that the limit should be 1. On the other hand we might argue that for x near to zero but not equal to zero we have $0 < \cos x < 1$, and any number between 0 and 1 when raised to a very large power is very small, so the limit should be zero. The problem with both these commonly used arguments is that each of them considers the two occurrences of x in the formula as if they were different variables, whereas they are of course changing at the same time.

In many algebraic problems involving indices, we find that taking logarithms transforms such an expression into one involving products and/or quotients. So let $y = (\cos x)^{\frac{1}{x}}$. Then $\ln y = \dfrac{\ln(\cos x)}{x}$, so we let $f(x) = \ln(\cos x)$ and $g(x) = x$. You should check that the conditions for l'Hôpital's Rule are satisfied. We then have $\dfrac{f'(x)}{g'(x)} = \dfrac{\frac{1}{\cos x}.(-\sin x)}{1} \to 0$ as $x \to 0$. Therefore by l'Hôpital's Rule $\ln y \to 0$ as $x \to 0$. Now the exponential function has a continuous graph, and so we deduce that $\exp(\ln y) \to \exp(0)$ as $x \to 0$, i.e., $y \to 1$ as $x \to 0$.

Example 2.28

Find $\lim\limits_{x \to 0} \dfrac{\tan x - x}{x - \sin x}$.

Let $f(x) = \tan x - x$, $g(x) = x - \sin x$. You should check that the conditions for l'Hôpital's Rule are satisfied. Now $f'(x) = \sec^2 x - 1$, $g'(x) = 1 - \cos x$. We see that in this case both derivatives tend to zero as $x \to 0$. So not only is $\dfrac{f(x)}{g(x)}$ a "$\frac{0}{0}$" type, but $\dfrac{f'(x)}{g'(x)}$ is also a "$\frac{0}{0}$" type. So we can try to apply l'Hôpital's Rule to $\dfrac{f'(x)}{g'(x)}$. Calculating the derivatives of the numerator and the denominator gives $f''(x) = 2\sec x.\sec x.\tan x$, $g''(x) = \sin x$. So we have

$$\frac{f''(x)}{g''(x)} = \frac{2\sec x.\sec x.\tan x}{\sin x} = \frac{2}{\cos^3 x} \to 2 \text{ as } x \to 0.$$

Therefore by l'Hôpital's Rule applied twice $\lim\limits_{x \to 0} \dfrac{\tan x - x}{x - \sin x} = 2$.

There are examples where l'Hôpital's Rule needs to be applied more than twice, but we must always check at each stage that the relevant quotient is a "$\frac{0}{0}$" type.

Example 2.29

Find $\lim\limits_{x \to \infty} \dfrac{x^2}{e^x}$.

We let $f(x) = x^2$ and $g(x) = e^x$. Then $f(x) \to \infty$ and $g(x) \to \infty$ as $x \to \infty$. So this satisfies the conditions for an "$\frac{\infty}{\infty}$" type. Differentiating then gives $f'(x) = 2x$ and $g'(x) = e^x$, so not only is $\dfrac{f(x)}{g(x)}$ an "$\frac{\infty}{\infty}$" type, but $\dfrac{f'(x)}{g'(x)}$ is also an "$\frac{\infty}{\infty}$" type. Differentiating again gives $f''(x) = 2$ and $g''(x) = e^x$. So we have

$$\frac{f''(x)}{g''(x)} = \frac{2}{e^x} \to 0 \text{ as } x \to \infty.$$

Therefore, applying l'Hôpital's Rule twice tells us that $\lim\limits_{x \to \infty} \dfrac{x^2}{e^x} = 0$.

This is an important result. It tells us that the exponential function grows more rapidly than x^2 as x tends to infinity. If we were to replace x^2 by x^n for any positive integer n we would obtain the same limit, zero, using l'Hôpital's Rule n times. So the exponential function grows faster than any power of x as x tends to infinity.

2.6 An Interesting Example

In this (optional) section we shall consider an example where trying to apply ordinary arithmetic rules leads to a contradiction. It indicates the need for care, and these kinds of problems are often studied in Real Analysis textbooks. In the 18th and 19th centuries it was assumed that properties of functions, such as continuity, were not changed under limiting operations, including those involved in infinite series. When this was finally shown not to be the case it caused significant mathematical problems, and several of the greatest mathematicians of the 19th century worked hard to resolve them, determining conditions under which such properties were preserved, and providing some understanding when they were not. The subject is discussed in Howie, Chapter 7.

We let

$$f(x) = \sum_{n=1}^{\infty} \frac{x^2}{(1+x^2)^n} = \frac{x^2}{(1+x^2)} + \frac{x^2}{(1+x^2)^2} + \frac{x^2}{(1+x^2)^3} + \cdots$$

$$= \frac{x^2}{(1+x^2)}\left[1 + \frac{1}{(1+x^2)} + \frac{1}{(1+x^2)^2} + \frac{1}{(1+x^2)^3} + \cdots\right].$$

The series in brackets is an infinite geometric series whose common ratio is $r = \dfrac{1}{(1+x^2)}$, and which therefore converges when x is non-zero. Using the well-known formula $a/(1-r)$ for the sum of an infinite geometric series with first term a and common ratio r gives, for $x \neq 0$,

$$f(x) = \frac{x^2}{1+x^2} \cdot \frac{1}{1 - \frac{1}{1+x^2}} = 1.$$

Now if we substitute $x = 0$ we get $f(0) = 0$. So the graph of $f(x)$ consists of the horizontal line $y = 1$ with a missing point at zero. There is a single isolated point at $(0,0)$. The function is discontinuous at 0. Furthermore we can see that $\lim_{x \to 0} f(x) = 1 \neq f(0)$. We can also see that for all values of n,

$\lim_{x \to 0} \dfrac{x^2}{(1+x^2)^n} = 0$. Therefore we can say that

$$\lim_{x \to 0} \sum_{n=1}^{\infty} \frac{x^2}{(1+x^2)^n} = 1 \ \text{ BUT } \ \sum_{n=1}^{\infty} \lim_{x \to 0} \frac{x^2}{(1+x^2)^n} = 0.$$

2.7 Limits using MAPLE

MAPLE has the capability to evaluate limits, and the syntax is straightforward. To find the limit of $f(x)$ as x tends to a we use the command

```
limit(f(x),x=a);
```

For example we can verify the result of Example 2.13 using the command

```
limit(x^2*sin(1/x),x=0);
```

and MAPLE will return the answer 0. What it will not do of course is to tell us how we could prove the result, or provide us with an understanding of why the result is true. When using MAPLE to find limits it is usually helpful therefore to plot the relevant function as well, to give a graphical insight.

MAPLE will also calculate limits as x tends to infinity. The result of Example 2.18 can be verified with the command

```
limit((3*x^2+4*x+4)/(4*x^2+3*x+2),x=infinity);
```

MAPLE returns the answer $\frac{3}{4}$ as an exact fraction, and not as a decimal.

Finally MAPLE will evaluate one-sided limits. We can investigate the floor function considered in Example 2.5, and verify that the left-sided limit as x tends to 2 is 1, whereas the right-sided limit is 2. The appropriate commands are

```
limit(floor(x),x=-1,left);
limit(floor(x),x=-1,right);
```

In this case the left- and right-sided limits are different, so if we use the command

```
limit(floor(x),x=-1);
```

MAPLE tells us that this limit is undefined.

2.8 Limits with Two Variables

In this (optional) section we begin to extend the ideas of limits to functions involving two real variables. This topic would be dealt with systematically in a course on Multivariate Calculus, and here we simply illustrate some possibilities by means of an example.

In considering limits such as $\lim_{x \to 0} f(x)$ we encountered the notion that x could approach zero from below or from above, and that sometimes the limiting behaviour was different in these two cases. For the purposes of graphs we take the variable x as being confined to the x-axis, a one-dimensional line. When we generalise these considerations to functions of two variables $f(x, y)$, then (x, y)

belongs to a two dimensional plane. We can consider limits such as

$$\lim_{(x,y)\to(0,0)} f(x,y).$$

in terms of (x, y) being "close to" $(0,0)$, and then investigating whether $f(x,y)$ is correspondingly "close to" $f(0,0)$. By analogy with one-sided limits we can then consider whether $f(x,y)$ has a limiting value if (x, y) in restricted in some way as it approaches $(0,0)$. This may be along some particular curve in the (x, y)-plane. The following example illustrates this idea.

Let $f(x,y) = \dfrac{x^2 - y^2}{x^2 + y^2}$, $(x, y) \neq (0,0)$. We consider what happens along several curves which approach $(0,0)$.

(a) Along the x-axis we have $y = 0$, and so

$$f(x,0) = \frac{x^2 - 0^2}{x^2 + 0^2} = 1.$$

So $f(x,0) \to 1$ as $x \to 0$.

(b) Along the y-axis we have $x = 0$, and so

$$f(0,y) = \frac{0^2 - y^2}{0^2 + y^2} = -1.$$

So $f(0,y) \to -1$ as $y \to 0$.

(c) Along the line $y = x$ we have

$$f(x,x) = \frac{x^2 - x^2}{x^2 + x^2} = 0.$$

So $f(x,x) \to 0$ as $x \to 0$.

(d) Along the line $y = 2x$ we have

$$f(x,2x) = \frac{x^2 - (2x)^2}{x^2 + (2x)^2} = -\frac{3}{5}.$$

So $f(x,2x) \to -\frac{3}{5}$ as $x \to 0$.

This suggests that $f(x,y)$ has different limiting values as $(x, y) \to (0,0)$ along various curves. Readers are invited to investigate what happens when $(x, y) \to (0,0)$ along a general line passing through $(0,0)$, along a parabola (such as $y = x^2$) passing through $(0,0)$, or along other curves passing through $(0,0)$.

Readers who are interested to see what a graphical representation looks like for the function we have considered can do so using the following MAPLE command.

```
plot3d((x^2-y^2)/(x^2+y^2),x=-1..1,y=-1..1);
```

EXERCISES

2.1. Plot the graph of the function given by the following formula, using a graphical calculator or MAPLE.

$$\frac{2x^2 + 5x + 7}{x^2 - 5x + 6}.$$

Describe the kinds of limiting behaviour you observe near the vertical asymptotes, and as $x \to \pm\infty$. You may need to use more than one plot to observe the various features of the graph, by varying the domain. In MAPLE the following commands should enable you to do this.

```
plot((2*x^2+5*x+7)/(x^2-5*x+6),x=0..5,y=-500..500);

plot((2*x^2+5*x+7)/(x^2-5*x+6),x=-1000..1000,y=1..3);
```

2.2. Plot the graphs of the functions given by the following formulae, using a graphical calculator or MAPLE. Describe the kinds of limiting behaviour you observe near any vertical asymptotes, and as $x \to \pm\infty$. Use the appropriate symbolic notation in each case. In some cases you will need more than one plot, as in the previous exercise. In cases involving roots you may need to use implicit plotting, as described in Section 1.3.

(a) $x^4 + 2x^2 - 3x + 1$;

(b) $3 + 2x^2 - x^5$;

(c) $\sqrt[3]{x^3 - 3x^2 - 6x + 8}$;

(d) $\dfrac{1}{\sqrt[3]{x^3 - 3x^2 - 6x + 8}}$;

(e) $\sqrt{x^2 + 4}$;

(f) $\dfrac{1}{\sqrt{x^2 + 4}}$;

(g) $\dfrac{x^2 - x - 2}{3x^2 + 4x + 5}$;

(h) $\dfrac{3x^2 + 4x + 5}{x^2 - x - 2}$;

(i) $\dfrac{x^2 - 2x + 3}{2x - 5}$;

(j) $\dfrac{2x - 3}{x^2 + 2x + 5}$;

(k) $\dfrac{2x + 5}{x^2 - 2x - 8}$;

(l) $\dfrac{1}{(\sin x + \cos x)^2}$;

(m) $\dfrac{\exp(2x)}{x^2}$;

(n) $\dfrac{1}{\exp(3x) - \exp(2x)}$;

(o) $\ln\left((x^2 - 4)^2\right)$;

(p) $\dfrac{\ln\left((x^2 - 4)^2\right)}{x}$;

(q) $\ln\left(2 + \sin\left(\dfrac{1}{x}\right)\right)$;

(r) $\ln\left(\dfrac{1}{x^2}\right)$;

(s) $\tan^{-1} x$

(t) $\tanh^{-1} x$.

2.3. Describe the limiting behaviour of the functions, given by the following formulae, on each side of the value of x indicated.

(a) $\exp\left(\dfrac{1}{x}\right)$, $\quad x = 0$;   (b) $\sqrt{\text{floor}(\sqrt{x})}$, $\quad x = 9$;

(c) $\dfrac{|x|}{x}$, $\quad x = 0$;   (d) $\dfrac{|\sin x|}{\sin x}$, $\quad x = \pi$;

(e) $\dfrac{\sqrt{x^2 - 2x + 1}}{x - 1}$, $\quad x = 1$;   (f) $\dfrac{\tan x}{|x|}$, $\quad x = 0$.

2.4. Find the following limits, using the appropriate algebraic rules.

(a) $\displaystyle\lim_{x \to 0} \dfrac{x^2 - 3x + 4}{\cos 2x}$;   (b) $\displaystyle\lim_{x \to \pi} \dfrac{2x - 3}{1 + \sin x}$;

(c) $\displaystyle\lim_{x \to 1} \sqrt{x^2 - 2x + 3}$;   (d) $\displaystyle\lim_{x \to -2} (x - 1)(x - 2)(x + 3)$;

(e) $\displaystyle\lim_{x \to 1} \dfrac{x^3 - 2}{x - 2}$;   (f) $\displaystyle\lim_{x \to -3} \dfrac{x^2 + 2x - 4}{x - 3}$;

(g) $\displaystyle\lim_{x \to 1} |2 - x - 3x^2| \cos(\pi x)$;   (h) $\displaystyle\lim_{x \to 0} \exp((x + 2) \sin x)$;

(i) $\displaystyle\lim_{x \to -1} \cosh(1 - x) \sin(\pi x)$;   (j) $\displaystyle\lim_{x \to \pi} \sqrt{(e^x + 3x - \ln x) \sin x}$.

2.5. Prove by squeezing that the value of each of the following limits is zero.

(a) $\displaystyle\lim_{x \to 0} |x| \sin\left(\dfrac{1}{x}\right)$;   (b) $\displaystyle\lim_{x \to \infty} e^{-x} \cos x$;

(c) $\displaystyle\lim_{x \to -\infty} e^x \sin(x^2 + 1)$;   (d) $\displaystyle\lim_{x \to \infty} (1 - \tanh x) \cos x$;

(e) $\displaystyle\lim_{x \to 1} |x - 1| \cos\left(\dfrac{1}{x - 1}\right)$;   (f) $\displaystyle\lim_{x \to \infty} \exp(\sin x - x)$.

2.6. The function $f(x)$ is bounded, i.e., there are constants A and B such that $A \le f(x) \le B$ for all x in the domain of f.

Show by squeezing that $\displaystyle\lim_{x \to 0} f(x) \sin x = 0$.

2.7. Use the method of Example 2.17 to determine the limiting behaviour at the vertical asymptotes of the function

$$f(x) = \dfrac{(x + 1)(x - 2)^2}{x(x - 1)(x + 2)^2}.$$

2.8. Evaluate the following limits, using algebraic manipulation.

(a) $\lim\limits_{x \to 2} \dfrac{x^2 - 4x + 4}{x^2 - 4}$;

(b) $\lim\limits_{x \to 1} \dfrac{x^2 + 6x - 7}{x^2 - 1}$;

(c) $\lim\limits_{x \to -1} \dfrac{x^3 + 1}{x + 1}$;

(d) $\lim\limits_{x \to 0} \dfrac{e^{2x} - 1}{e^x - 1}$;

(e) $\lim\limits_{x \to 2} \left(\dfrac{4}{x^2 - 4} - \dfrac{1}{x - 2} \right)$;

(f) $\lim\limits_{x \to 1} \dfrac{x^2 - 1}{\sqrt{x} - 1}$;

(g) $\lim\limits_{x \to -1} \dfrac{x^2 - 1}{\sqrt{5 + x} - 2}$;

(h) $\lim\limits_{x \to 0} \dfrac{\sqrt{1 - 2x^2} - \sqrt{1 + 2x^2}}{x^2}$;

(i) $\lim\limits_{x \to \infty} \dfrac{x^4 + 2x^2 + 1}{2x^4 - 3x^3 + x}$;

(j) $\lim\limits_{x \to \infty} \dfrac{x^2 + 2x + 1}{x^3 - x^2 - x - 1}$;

(k) $\lim\limits_{x \to -\infty} \dfrac{x^3 + x^2 - x - 1}{2x + 3}$;

(l) $\lim\limits_{x \to -\infty} \dfrac{x^2 + 2\sqrt{x} + \sqrt[3]{x}}{\sqrt{x} - 1 - x^2}$;

(m) $\lim\limits_{x \to -\infty} \dfrac{\sqrt{2x^2 + 3x + 5}}{x - 4}$;

(n) $\lim\limits_{x \to -\infty} \dfrac{|2x - 5|}{3x + 1}$;

(o) $\lim\limits_{x \to \infty} \dfrac{x^2 + x + \cos^2 x}{2x^2 - \sin^2 2x}$;

(p) $\lim\limits_{x \to \infty} \dfrac{e^x - \sin x + 1}{2e^x + \cos x - 3}$.

2.9. Use the method of change of variable to evaluate the following limits.

(a) $\lim\limits_{x \to 0} \dfrac{e^{2x} - 1}{e^x - 1}$;

(b) $\lim\limits_{x \to 1} \dfrac{\sin(\ln x)}{\ln x}$;

(c) $\lim\limits_{x \to 0} \dfrac{\sqrt{2x}}{\sin(2\sqrt{x})}$;

(d) $\lim\limits_{x \to e} \dfrac{(\ln x)^2 - 1}{\ln x - 1}$;

(e) $\lim\limits_{x \to 0} \dfrac{x}{\sin^{-1} x}$;

(f) $\lim\limits_{x \to -\infty} \dfrac{\sin^{-1}(e^x)}{e^x}$.

2.10. Use l'Hôpital's rule to find the values of the following limits.

(a) $\lim\limits_{x \to 1} \dfrac{\ln x}{x^2 - 1}$;

(b) $\lim\limits_{x \to \pi} \dfrac{\sin x}{x - \pi}$;

(c) $\lim\limits_{x \to 0} (\operatorname{cosec} x - \cot x)$;

(d) $\lim\limits_{x \to 0} \left(\dfrac{1}{\sin x} - \dfrac{1}{x} \right)$.

(e) $\lim\limits_{x \to 0} \dfrac{\cos ax - 1}{\cos bx - 1}$ $(b \neq 0)$;

(f) $\lim\limits_{x \to 0} \left(\dfrac{1}{x^2} - \dfrac{\cos(ax)}{x^2} \right)$;

(g) $\lim\limits_{x \to 0} \dfrac{\sin^3 x}{x - \tan x}$.

(h) $\lim\limits_{x \to 0} \dfrac{\sin x - x + \frac{x^3}{6}}{x^5}$;

(i) $\lim\limits_{x \to 0^+} x^{\sin(x)}$;

(j) $\lim\limits_{x \to 0^+} \left(\sqrt[3]{x} \right)^x$;

(k) $\lim\limits_{x \to \infty} \left(1 + \dfrac{2}{x} \right)^x$;

(l) $\lim\limits_{x \to 0} (1 + \sin x)^{\frac{1}{x}}$.

2.11. Find examples of pairs of functions $f(x), g(x)$ satisfying the following as $x \to \infty$.

(a) $f(x) \to \infty, g(x) \to \infty$ and $f(x) - g(x) \to 0$;

(b) $f(x) \to \infty, g(x) \to \infty$ and $f(x) - g(x) \to \infty$;

(c) $f(x) \to \infty, g(x) \to \infty$ and $f(x) - g(x) \to -\infty$;

(d) $f(x) \to \infty, g(x) \to \infty$ and $f(x) - g(x) \to 3$.

2.12. The function $f(x)$ is defined by $\dfrac{|x+k|}{x^2 - k^2}$ $\quad (k \neq 0)$.

What is the domain of $f(x)$? Sketch the graph of $f(x)$ showing the discontinuities. (You can use MAPLE to get an idea of what the graph looks like.) Use the graph to find the values of the left- and right-sided limits at each of the discontinuities, and show how to derive these limits from the formula for $f(x)$.

2.13. Investigate the behaviour of the function of two variables defined by

$$f(x, y) = \frac{x^2 - y^3}{x^3 - y^2}$$

as $(x, y) \to (0, 0)$ along various curves passing through $(0, 0)$, along the lines of Section 2.8.

3
Differentiation

The differential calculus has two major areas of use and origin. One is geometry, and the problem of finding tangents to curves. The other is motion (speed, velocity, acceleration) and other rates of change. Both of these lead to the definition of the derivative in terms of a limit.

3.1 The Limit Definition

We shall explore the definition of derivative by considering the problem of finding the gradient of a curve, and therefore its tangent.

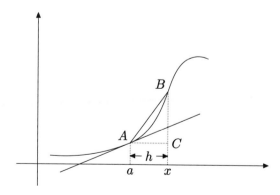

Figure 3.1   Chord slope diagram

Figure 3.1 represents the graph of a function $y = f(x)$. The tangent line at the point A can be considered as the limiting position of the chord AB as B tends towards A. This is achieved by letting x tend to a, or equivalently by letting h tend to 0, since $x = a + h$. It is important to note that this is not just a one-sided limit, and so a diagram where B is to the left of A, where h is negative, is equally valid. The gradient of the tangent will therefore be the limit as x tends to a of the gradient of the chord. The coordinates of the points labelled in Figure 3.1 are as follows.

$$A(a, f(a)), \; B(x, f(x)), \; C(x, f(a)).$$

Therefore the gradient (slope) of the chord is given by

$$\frac{BC}{AC} = \frac{f(x) - f(a)}{x - a} = \frac{f(a + h) - f(a)}{h}.$$

Taking limits therefore gives the gradient of the tangent, giving rise to the following definition.

Definition 3.1

The function $f(x)$ whose domain includes some interval containing the point a is said to be **differentiable** at a if the following limit exists.

$$\lim_{x \to a} \frac{f(x) - f(a)}{x - a}.$$

The value of this limit is called the **derivative** (or **differential coefficient**) of f at a, denoted by $f'(a)$. We therefore have

$$f'(a) = \lim_{x \to a} \frac{f(x) - f(a)}{x - a} = \lim_{h \to 0} \frac{f(a + h) - f(a)}{h}.$$

In many cases a function will be differentiable for all (or most) of the values of x in the domain. In that case we think of a as a variable and use the term **derivative** for the function whose value at $x = a$ is $f'(a)$.

In fact a variety of terminology is encountered in this topic. Terms such as differential coefficient, derived function, differential and derivative are all used, sometimes to convey different shades of meaning and interpretation. In an introductory account such distinctions are not so important, whereas they are in more advanced areas of calculus. We shall use the term derivative to refer to the function resulting from the process of differentiation (sometimes called the **derived function**), and also to the value of this derived function at some point of its domain.

There are two types of notation in common use, the dash notation $f'(x)$, $f''(x)$, etc., and the Leibniz notation $\dfrac{dy}{dx}, \dfrac{d^2y}{dx^2}$, etc. Readers will have encountered these in school calculus. Sometimes the Leibniz notation is more helpful than the dash notation, and vice-versa, and we shall use them according to this criterion.

3.2 Using the Limit Definition

In this section we consider some examples where we can find the derivative directly from the limit, together with an example where the derivative does not exist. We also prove a basic property of derivatives useful in graph sketching. In practice we do not rely heavily on the limit definition. Instead we use algebraic rules for differentiation and apply them to functions whose derivatives we already know. This mirrors the procedure we used in Chapter 2 to find limits, and the processes of integration we shall develop in later chapters. The first two examples show that the basic derivatives can be found using the limit definition. This is sometimes referred to as "finding the derivative from first principles", the first principles in question being the limit definition.

Example 3.2

Use the limit definition to find the derivative of $f(x) = x^2$.

Applying the limit definition gives

$$f'(a) = \lim_{x \to a} \frac{f(x) - f(a)}{x - a} = \lim_{x \to a} \frac{x^2 - a^2}{x - a} = \lim_{x \to a} (x + a) = 2a.$$

Example 3.3

Use the limit definition to find the derivative of $f(x) = \sin x$.

$$
\begin{aligned}
f'(a) &= \lim_{h \to 0} \frac{f(a+h) - f(a)}{h} = \lim_{h \to 0} \frac{\sin(a+h) - \sin a}{h} \\
&= \lim_{h \to 0} \frac{2\cos\left(a + \frac{h}{2}\right)\sin\left(\frac{h}{2}\right)}{h} = \lim_{h \to 0} \cos\left(a + \frac{h}{2}\right)\frac{\sin\left(\frac{h}{2}\right)}{\frac{h}{2}} \\
&= \cos(a).1 = \cos a.
\end{aligned}
$$

Here we have used the limit obtained in Example 2.2, with $x = \frac{h}{2}$.

Example 3.4

Show that $f(x) = |x|$ is not differentiable at 0.

We recall the graph of $y = |x|$ shown in Figure 1.8, and notice that it has a sharp corner at $x = 0$. The gradient to the right is 1 and the gradient to the left is -1, indicating that the gradient at 0 cannot be well-defined. The limit definition confirms this, as follows.

$$\frac{f(0+h) - f(0)}{h} = \begin{cases} \frac{h}{h} = 1 & \text{for } h > 0, \\ \frac{-h}{h} = -1 & \text{for } h < 0. \end{cases}$$

So the left- and right-sided limits are different. Therefore the (two-sided) limit does not exist, and so the modulus function is not differentiable at 0.

Example 3.5

In Section 1.8 we considered functions defined in pieces. In that section we introduced the function

$$k(x) = \begin{cases} x^2 & \text{if } x \geq 0, \\ x^3 & \text{if } x < 0. \end{cases}$$

If $x > 0$ then $k(x) = x^2$ and so $k'(x) = 2x$. If $x < 0$, $k(x) = x^3$ and so $k'(x) = 3x^2$. But to investigate differentiability at $x = 0$ we need to use the limit definition, as follows.

$$\frac{k(0+h) - k(0)}{h} = \begin{cases} \frac{h^2 - 0}{h} = h & \text{if } x > 0, \\ \frac{h^3 - 0}{h} = h^2 & \text{if } x < 0. \end{cases}$$

We conclude from this that

$$\lim_{h \to 0+} \frac{k(0+h) - k(0)}{h} = \lim_{h \to 0+} h = 0;$$

$$\lim_{h \to 0-} \frac{k(0+h) - k(0)}{h} = \lim_{h \to 0-} h^2 = 0.$$

The left- and right-hand limits are equal, so

$$\lim_{h \to 0} \frac{k(0+h) - k(0)}{h} = 0.$$

Therefore $k(x)$ is differentiable at $x = 0$ and $k'(x) = 0$.

Figure 3.2 shows the graph of $k'(x)$. We can see that there appears to be a sharp corner at $x = 0$, as there is for $|x|$. This suggests that $k'(x)$ is not

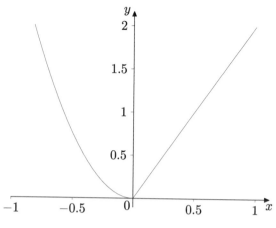

Figure 3.2   A piecewise derivative

differentiable at $x = 0$, and we can prove this using the limit definition. We have

$$\frac{k'(0+h) - k'(0)}{h} = \begin{cases} \frac{2h-0}{h} = 2 & \text{if } x > 0, \\ \frac{3h^2-0}{h} = 3h & \text{if } x < 0. \end{cases}$$

So we deduce that

$$\lim_{h \to 0^+} \frac{k'(0+h) - k'(0)}{h} = \lim_{h \to 0^+} 2 = 2;$$

$$\lim_{h \to 0^-} \frac{k'(0+h) - k'(0)}{h} = \lim_{h \to 0^-} 3h = 0.$$

The left-and right-hand limits are not the same, and so

$$\lim_{h \to 0} \frac{k'(0+h) - k'(0)}{h} \quad \text{does not exist.}$$

This shows that $k'(x)$ is not differentiable at $x = 0$.

Example 3.6

Prove that if a differentiable function is increasing (see Definition 1.35) then its derivative is non-negative.

Suppose that for all a, b in the domain of f satisfying $a \leq b$, we have $f(a) \leq f(b)$. Let x denote an arbitrary number in the domain of f. Then if $h > 0$ we have $\frac{f(x+h) - f(x)}{h} \geq 0$, because both numerator and denominator are positive (or zero). If $h < 0$ we also have $\frac{f(x+h) - f(x)}{h} \geq 0$, because in this case both numerator and denominator are negative (or zero). Therefore we must have

$$f'(x) = \lim_{h \to 0} \frac{f(x+h) - f(x)}{h} \geq 0.$$

We can prove in a very similar fashion that if a differentiable function is decreasing then its derivative is non-positive. Note that if a function is strictly decreasing this does not imply that its derivative is strictly positive everywhere. For example $f(x) = x^3$ is strictly increasing, but $f'(0) = 0$.

We shall prove a converse of this result under appropriate conditions in Section 6.2.

3.3 Basic Rules of Differentiation

The basic algebraic rules of differentiation enable us to differentiate sums, products and quotients of functions whose derivatives we already know. We assume that the derivatives in the table below are known from school mathematics.

$f(x)$	$f'(x)$
x^n	nx^{n-1}
$\sin x$	$\cos x$
$\cos x$	$-\sin x$
$\tan x$	$\sec^2 x$
e^x	e^x
$\ln x$	$\frac{1}{x}$

The basic rules of differentiation are summarised as follows.

Suppose that f and g are differentiable functions. Then for any constants $A, B,$

$$\frac{d}{dx}(Af(x) + Bg(x)) = Af'(x) + Bg'(x) \quad \text{(the sum rule)};$$

$$\frac{d}{dx}(f(x)g(x)) = f(x)g'(x) + g(x)f'(x) \quad \text{(the product rule)};$$

$$\frac{d}{dx}\left(\frac{f(x)}{g(x)}\right) = \frac{g(x)f'(x) - f(x)g'(x)}{(g(x))^2} \quad (g(x) \neq 0) \quad \text{(the quotient rule)}.$$

This is a case where it is convenient to use both notations for derivatives together.

Example 3.7

We prove the product rule from the limit definition of the derivative.

In the second line of the proof below, we have introduced an additional term in the numerator, together with its negative, hence preserving equality. Its purpose is to enable us to analyse separately the changes in f and the changes in g as h tends to zero. This is apparent in the third line, where we can see the chord-slope quotients for f and for g, whose limits are the respective derivatives as h tends to zero.

$$\frac{f(x+h)g(x+h) - f(x)g(x)}{h}$$

$$= \frac{f(x+h)g(x+h) - f(x+h)g(x) + f(x+h)g(x) - f(x)g(x)}{h}$$

$$= f(x+h)\frac{g(x+h) - g(x)}{h} + g(x)\frac{f(x+h) - f(x)}{h}$$

$$\to f(x)g'(x) + g(x)f'(x) \quad \text{as} \quad h \to 0.$$

3.4 The Chain Rule

The Chain Rule (or function of a function rule) tells us how to differentiate composite functions, and while it is usually part of school calculus, it is sufficiently important to merit some revision. The rule is stated as follows.

Suppose that the function g is differentiable at x, and that the function f is differentiable at $g(x)$. Then the derivative of $f \circ g(x) = f(g(x))$ is given by $(f \circ g)'(x) = f'(g(x))g'(x)$.

The rule can be stated using the Leibniz notation as follows.

If $y = f(u)$ and $u = g(x)$ then $\dfrac{dy}{dx} = \dfrac{dy}{du}\dfrac{du}{dx}$.

To derive the chain rule from the limit definition we proceed as follows.

$$\frac{f(g(x+h)) - f(g(x))}{h} = \frac{f(g(x+h)) - f(g(x))}{g(x+h) - g(x)}\frac{g(x+h) - g(x)}{h}.$$

Here we have introduced the term $g(x+h) - g(x)$ in the numerator and in the denominator. This helps to separate the behaviour of f and that of g. We then let $g(x) = u$ and $g(x+h) = u + k$. Then $k \to 0$ as $h \to 0$. We therefore have

$$\frac{f(g(x+h)) - f(g(x))}{h} = \frac{f(u+k) - f(u)}{k}\frac{g(x+h) - g(x)}{h}$$

$$\to f'(u)g'(x) = f'(g(x))g'(x) \quad \text{as} \quad h \to 0.$$

This argument appears to be sound, but in fact there is a problem at the beginning, for we cannot be sure that $g(x+h) - g(x)$ will not be zero for some values of h arbitrarily close to 0, and we cannot divide by zero. The argument does however provide an intuitive justification relating the chain rule to the limit definition. A properly rigorous proof is given in Howie Chapter 4.

The use of an "intermediate variable" such as u in applying the chain rule is often helpful and we shall employ it in the following examples.

Example 3.8

Differentiate $\ln(\cos x)$.

Using the Chain Rule, let $y = \ln u$, $u = \cos x$. The derivative is given by

$$\frac{dy}{dx} = \frac{dy}{du}\frac{du}{dx} = \frac{1}{u}(-\sin x) = \frac{-\sin x}{\cos x} = -\tan x.$$

Example 3.9

Differentiate a^x with respect to x.

A common mistake is to assume that we use the simple formula for powers and write the derivative as xa^{x-1}. This is WRONG. What has been calculated here is the derivative with respect to a, not the derivative with respect to x. To do the calculation correctly, we recall from Definition 1.27 that $a^x = e^{x\ln a}$. We therefore use the chain rule, letting $y = e^u$ and $u = x\ln a$. This gives

$$\frac{dy}{dx} = \frac{dy}{du}\frac{du}{dx} = e^u \ln a = e^{x\ln a}\ln a = a^x \ln a.$$

Example 3.10

Differentiate $\sin(\ln(x^3 - 4x))$.

In this example we have repeated composition, and we extend the chain rule using two intermediate variables.

We let $y = \sin u$, $u = \ln t$, $t = x^3 - 4x$. The derivative is then given by

$$\begin{aligned}\frac{dy}{dx} &= \frac{dy}{du}\frac{du}{dt}\frac{dt}{dx} = (\cos u).\frac{1}{t}.(3x^2 - 4)\\ &= \frac{\cos\left(\ln(x^3 - 4x)\right)(3x^2 - 4)}{x^3 - 4x}.\end{aligned}$$

Example 3.11

Differentiate $\ln\left(\tan\left(2 + x^4\right)^{\frac{1}{2}}\right)$.

There is no limit to the number of stages of composition.

In this case we introduce the requisite number of intermediate variables, letting $y = \ln p$, $p = \tan q$, $q = r^{\frac{1}{2}}$, $r = 2 + x^4$.

Applying the chain rule therefore gives

$$\frac{dy}{dx} = \frac{dy}{dp}\frac{dp}{dq}\frac{dq}{dr}\frac{dr}{dx} = \frac{1}{p}.\sec^2 q.\frac{1}{2}r^{-\frac{1}{2}}.4x^3$$

$$= \frac{\sec^2\left((2+x^4)^{\frac{1}{2}}\right).2x^3}{\tan\left((2+x^4)^{\frac{1}{2}}\right)(2+x^4)^{\frac{1}{2}}}.$$

Example 3.12

Differentiate $f(x) = x^2 \cos\left(\dfrac{1}{x}\right)$.

In examples such as this one we have to use more than one of the rules. Firstly we need the product rule since the function is x^2 multiplied by a cosine term. Secondly the cosine terms itself is composite, and so we need the chain rule. So applying both rules gives

$$f'(x) = 2x\cos\left(\frac{1}{x}\right) + x^2\left(-\sin\left(\frac{1}{x}\right)\right)\left(-\frac{1}{x^2}\right) = 2x\cos\left(\frac{1}{x}\right) + \sin\left(\frac{1}{x}\right).$$

Figure 3.3 shows a MAPLE plot of this formula for $f'(x)$. (We pointed out the limitations of MAPLE plots of such functions in Example 2.3)

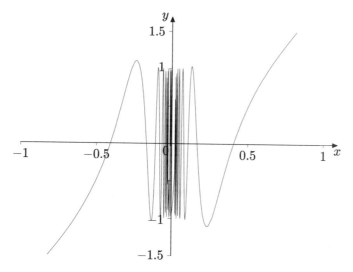

Figure 3.3   A discontinuous derivative

Now this calculation is not valid when $x = 0$, and indeed $f(x)$ is not defined there. However, the squeezing argument used in Example 2.13 shows that $f(x)$ has the limit zero as x tends to zero, and that if we extend the definition by letting $f(0) = 0$ the resulting function is continuous at 0. So what about differentiability? We can't substitute $x = 0$ in the formula we have just found, so we have to go back to the limit definition, and investigate whether the appropriate limit exists.

$$\frac{f(0+h) - f(0)}{h} = \frac{h^2 \cos\left(\frac{1}{h}\right)}{h} = h \cos\left(\frac{1}{h}\right) \to 0 \ \text{ as } \ h \to 0,$$

by the squeezing argument. Therefore f is differentiable at 0 and $f'(0) = 0$. This is a very interesting example, because although f is differentiable at 0, we can see that the formula for $f'(x)$ does not have a limit as $x \to 0$, because the sin term oscillates infinitely often in any interval containing $x = 0$, as we established in Example 2.3. So f is differentiable everywhere but the derivative is discontinuous at $x = 0$.

3.5 Higher Derivatives

If we have a function $y = f(x)$ specified by a given formula and we differentiate it we obtain the formula for $f'(x)$, which we can usually differentiate again, and in many cases we can repeat the process several times. This gives a sequence of derivatives, denoted by

$$f'(x), f''(x), f'''(x), f^{(4)}(x), \ldots, f^{(n)}(x), \ldots,$$

or, using the Leibniz notation for derivatives,

$$\frac{dy}{dx}, \frac{d^2y}{dx^2}, \frac{d^3y}{dx^3}, \ldots, \frac{d^ny}{dx^n}, \ldots.$$

Higher derivatives have applications, for example in mechanics where the second derivative of position relates to acceleration, and in coordinate geometry as we shall see in Chapter 5.

Example 3.13

Find the n-th derivative of $f(x) = \ln(2x + 3)$.

Calculating the first few derivatives, using the chain rule, is relatively straightforward, giving

$$f(x) \quad = \quad \ln(2x + 3);$$

$$f'(x) = \frac{2}{2x+3};$$

$$f''(x) = \frac{-4}{(2x+3)^2};$$

$$f'''(x) = 2.\frac{8}{(2x+3)^3};$$

$$f^{(4)}(x) = 3.2.\frac{-16}{(2x+3)^4};$$

$$f^{(5)}(x) = 4.3.2.\frac{32}{(2x+3)^5}.$$

There is a clear pattern appearing here, which enables us to conjecture a formula for the n-th derivative, namely

$$f^{(n)}(x) = (n-1)!\frac{(-1)^{(n+1)}2^n}{(2x+3)^n}.$$

To prove this we need to use the method of mathematical induction. Readers who are not familiar with this method can normally rely on conjecturing such results by generalising, without proof, the pattern observed in the first few cases. The details of the method are discussed in Howie Chapter 1. The inductive proof is as follows, for readers who are familiar this style of proof.

The case $n = 1$ has already been established. If the result were true for $n = k$ then

$$f^{(k)}(x) = (k-1)!\frac{(-1)^{(k+1)}2^k}{(2x+3)^k} = (k-1)!(-1)^{(k+1)}2^k(2x+3)^{-k}.$$

Differentiating this formula once more would give

$$f^{(k+1)}(x) = (k-1)!(-1)^{(k+1)}2^k(-k)(2x+2)^{-k-1}.2 = k!\frac{(-1)^{(k+2)}2^{k+1}}{(2x+3)^{k+1}},$$

giving the result for $n = k+1$, so proving the general result by induction.

Example 3.14

Find successive derivatives for $f(x) = \sin\left(x^2\right)$.

Using the chain rule and the product rule gives

$$f(x) = \sin\left(x^2\right);$$

$$f'(x) = 2x\cos\left(x^2\right);$$

$$f''(x) = -4x^2\sin\left(x^2\right) + 2\cos\left(x^2\right).$$

It can be seen even at this stage that the expressions are increasing in complexity, so that for example the sixth derivative is given by

$$f^{(6)}(x) = -64x^6 \sin\left(x^2\right) + 480x^4 \cos\left(x^2\right) + 720x^2 \sin\left(x^2\right) - 120 \cos\left(x^2\right).$$

While it is clear that there is a pattern here, it is not easy to formulate an expression for the n-th derivative.

3.6 Differentiation using MAPLE

The MAPLE command for differentiation is straightforward. For example executing the command `diff(sin(x^2),x);` will give the derivative of $\sin\left(x^2\right)$. The role of the x at the end of the command it important. An error message will result if it is omitted. It tells MAPLE what the variable of differentiation is. This can be illustrated with the two commands
    `diff(x^2*y^3,x);` and `diff(x^2*y^3,y);`
which will return the outputs $2xy^3$ and $3x^2y^2$ respectively.

It should be noted that the output from MAPLE will not necessarily look identical to a formula we would obtain "by hand". For example the command `diff(tan(x),x);` produces $1 + \tan^2 x$ instead of $\sec^2 x$, but the two are of course identical. The command `simplify(%);` will sometimes transform an answer into a more recognisable form. The percentage (%) symbol denotes "the previous expression" or "the output of the previous calculation", and using it can save having to repeat typing in a complicated expression.

Calculating successive derivatives is also straightforward using MAPLE. One can enter a formula such as `sin(x^2);` Executing this command simply prints the formula $\sin\left(x^2\right)$ on the screen. The command `diff(%,x);` then calculates the derivative with respect to x. Following this with the same command `diff(%,x);` will therefore repeat the process, giving the second derivative. This procedure could be used to calculate the sixth derivative quoted in Example 3.14. Alternatively we could calculate the sixth derivative directly using the command
    `diff(sin(x^2),x$6);`
where the $6 sign tells MAPLE that we want the sixth derivative.

Finally we note that one need not have a particular formula, so that for example MAPLE will help us if we forget the product rule. Entering the command `diff(f(x)*g(x),x);` produces the output

$$\left(\frac{d}{dx}f(x)\right)g(x) + f(x)\left(\frac{d}{dx}g(x)\right).$$

EXERCISES

3.1. Use the limit definition to find the derivative of each of the following functions

$$\text{(a) } x^3; \quad \text{(b) } x^{-1}; \quad \text{(c) } \cos x; \quad \text{(d) } \tan x; \quad \text{(e) } e^x.$$

3.2. Use the sum, product and quotient rules to find the derivative of each of the functions defined by the following expressions.

(a) $8x^{3/4}$;   (b) $\sinh x$;   (c) $e^v \sin v$;

(d) $x^2 \tan x$;   (e) $t \sin t + \cos t$;   (f) $\tanh x$;

(g) $\dfrac{3x - 2}{2x - 3}$;   (h) $\dfrac{t^2 + 2t}{t^2 - 1}$;   (i) $\dfrac{1 - 4x}{x^{2/3}}$;

(j) $\dfrac{\cos x}{1 + 2\sin x}$;   (k) $\dfrac{e^w}{1 - \tan w}$;   (l) $\dfrac{e^x \ln x}{x^2 + 2x^3}$.

3.3. Use the chain rule to find the derivative of each of the functions defined by the following expressions.

(a) $\cos\left(\sqrt{x}\right)$;   (b) $\cosh(\cos t)$;   (c) 2^{-x};

(d) $\ln(\ln(\ln x))$;   (e) $\left(1 + s^{2/3}\right)^{3/2}$;   (f) $\left(3 - 2t^2\right)^{-3/4}$;

(g) $\tan\left(\tfrac{1}{x}\right)$;   (h) $\sqrt{\sin\left(v^2\right)}$;   (i) $\sin(2\cos 3x)$;

(j) 3^{3^x};   (k) $\cos(\ln x)$;   (l) $\sqrt[3]{\ln t}$.

3.4. Find the derivative, with respect to x, of each of the functions defined by the following expressions, using the appropriate combinations of rules.

(a) $\ln(x \sin x)$;   (b) $\sin\left(\dfrac{x}{\cos x}\right)$;

(c) $\sqrt{x + e^x}$;   (d) $\sqrt{x}\sqrt[3]{1 + x^2}$;

(e) $\cosh(x \ln x)$;   (f) $\dfrac{\sin\left(x^2\right)}{\sec\left(x^2\right)}$;

(g) $\tan(3x^3)\cot(3x^3)$;   (h) $\tan\left(a^2\left(1 + x^2\right)\right)$;

(i) $2^{x \sin x}$;   (j) $a\sin(bx) + b\sin(ax)$;

(k) $\left(x^2 \ln x\right)^{\left(b^2\right)}$;   (l) $\tan^2\left(\dfrac{1}{cx^2 + d}\right)$.

3.5. Find successive derivatives of $f(x) = \sin(2x - 5)$. On the basis of the first few derivatives write down a general formula for $f^{(2n)}(x)$ and for $f^{(2n+1)}(x)$. Prove these results by the method of mathematical induction.

Challenge: find a single formula which covers both the separate formulae above.

3.6. Use MAPLE to find successive derivatives of $f(x) = e^{ax} \sin ax$. Write down general formulae for $f^{(4n)}(x)$, $f^{(4n+1)}(x)$, $f^{(4n+2)}(x)$ and $f^{(4n+3)}(x)$. Prove these results by the method of mathematical induction.

3.7. The derivative of an even function is an odd function.

The derivative of an odd function is an even function.

(a) Write a clear explanation of these results based on diagrams.

(b) Prove the results by differentiating the equations which define an even function and an odd function, given in Definitions 1.4 and 1.5.

(c) Prove the results from the limit definition of the derivative, given in Definition 3.1.

Do you think the converses are true, namely that every odd (even) function is the derivative of an even (odd) function? If you think so, give a proof. If you do not think so, give a counter-example. If you think the converses are true only for some kinds of function, describe such a set of functions and prove the converses for this set.

4

Techniques of Differentiation

In this chapter we shall explore some techniques of differentiation which deal with functions specified in various forms. We shall consider functions defined implicitly, functions defined parametrically, functions involving powers, and inverse functions. We shall also discuss Leibniz Theorem, a result which enables higher derivatives of products to be calculated.

4.1 Implicit Differentiation

Sometimes we are not given y as a function of x explicitly, but instead have an equation connecting them which we may be unable to solve explicitly for either x or y. We may still want to find $\dfrac{dy}{dx}$, but we shall find that the resulting expression still involves both variables.

The following example illustrates what is meant.

Example 4.1

Find the gradient $\dfrac{dy}{dx}$ at the point $(1, 2)$ on the curve whose equation is

$$x^3 - 5xy^2 + y^3 + 11 = 0.$$

Figure 4.1 shows that the curve is not the graph of y as a function of x. Indeed when $x = 1$ there are three possible values of y on the part of the graph

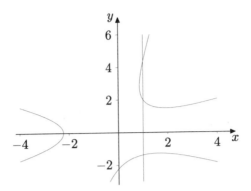

Figure 4.1   Graph of $x^3 - 5xy^2 + y^3 + 11 = 0$

shown. This is indicated by the inclusion of the line $x = 1$ in Figure 4.1. One of the three intersections of this line with the graph is of course the given point $(1, 2)$. If we consider a small part of the curve in the neighbourhood of that point then it is the graph of a function $y = y(x)$ which is one of the solutions of the equation of the curve and which specifies part of the graph near to $(1, 2)$. We cannot find $y(x)$ explicitly in terms of x, otherwise we would be able to use the normal procedures of differentiation.

The function $y(x)$ satisfies the equation of the curve, namely

$$x^3 - 5x(y(x))^2 + (y(x))^3 + 11 = 0.$$

We therefore have to use the chain rule to differentiate the y^2 and y^3 terms, and the product rule for the second term, involving x and y. Using the chain rule for the terms involving powers of $y(x)$ gives

$$\frac{d}{dx}\left(y(x)^3\right) = 3\left(y(x)\right)^2 \frac{dy}{dx};$$
$$\frac{d}{dx}\left(y(x)^2\right) = 2y(x)\frac{dy}{dx}.$$

Differentiating the equation of the curve with respect to x therefore gives

$$3x^2 - 5\left(x.2y(x)\frac{dy}{dx} + (y(x))^2\right) + 3\left(y(x)\right)^2 \frac{dy}{dx} = 0.$$

Rearranging this gives

$$\left(3(y(x))^2 - 10xy(x)\right)\frac{dy}{dx} = 5(y(x))^2 - 3x^2,$$

and therefore
$$\frac{dy}{dx} = \frac{5(y(x))^2 - 3x^2}{3(y(x))^2 - 10xy(x)} = \frac{5y^2 - 3x^2}{3y^2 - 10xy}.$$

The gradient at the point $(1, 2)$ is then found by substituting these values for x and y in this expression, giving $-17/8$. This value is consistent with Figure 4.1, where the tangent line at $(1, 2)$ does indeed appear to have a fairly steep negative gradient.

When we get used to this procedure we do not normally write $y(x)$ in full, but just use y throughout, as in the next example.

Example 4.2

Given $\cos(xy) = \exp(x + y)$, find $\dfrac{dy}{dx}$ in terms of x and y.

This is a purely algebraic problem. We first apply the chain rule to both sides, giving

$$-\sin(xy)\frac{d}{dx}(xy) = \exp(x + y)\frac{d}{dx}(x + y).$$

The left hand side needs the product rule, and applying this gives

$$-\sin(xy)\left(y + x\frac{dy}{dx}\right) = \exp(x + y)\left(1 + \frac{dy}{dx}\right).$$

We now collect all the terms involving the derivative and then divide to isolate the derivative, as we did in Example 4.1. We then obtain

$$\frac{dy}{dx} = -\frac{\exp(x + y) + y\sin(xy)}{x\sin(xy) + \exp(x + y)},$$

provided that the denominator is not zero.

Note that substituting arbitrary values of x and y in this equation is meaningless. The point (x, y) would have to satisfy the original equation in order that $\dfrac{dy}{dx}$ could be interpreted as the gradient of the curve.

Example 4.3

Given $xy + e^y = 0$, find $\dfrac{dy}{dx}$ and $\dfrac{d^2y}{dx^2}$ in terms of x and y.

Differentiating the equation with respect to x gives

$$y + x\frac{dy}{dx} + e^y\frac{dy}{dx} = 0.$$

We could solve this to find the derivative, and then differentiate the resulting equation. Instead we differentiate once more without rearranging first, giving

$$\frac{dy}{dx} + \frac{dy}{dx} + x\frac{d^2y}{dx^2} + e^y\frac{dy}{dx}\cdot\frac{dy}{dx} + e^y\frac{d^2y}{dx^2} = 0.$$

We now rearrange the two equations to give

$$\frac{dy}{dx} = -\frac{y}{x + e^y};$$

$$\frac{d^2y}{dx^2} = -\frac{2\frac{dy}{dx} + e^y \left(\frac{dy}{dx}\right)^2}{x + e^y}$$

$$= -\frac{-\frac{2y}{x+e^y} + e^y \left(\frac{-y}{x+e^y}\right)^2}{x + e^y}$$

$$= -\frac{-2y(x + e^y) + y^2 e^y}{(x + e^y)^3}.$$

One can imagine that if the initial equation were more complicated then finding the second derivative would be very involved, and so it is useful to see how MAPLE could tackle the calculations. We might think that we could undertake the first step in the calculations above using the command `diff(x*y+exp(y),x);` Unfortunately this just returns the output y, because MAPLE does not know that y is meant to be a function of x. We must use $y(x)$ in place of y, as we did in the first example in this section. The following sequence of commands can be used to solve the problem.

        diff(x*y(x)+exp(y(x)),x);
        diff(%,x);

We now rearrange this equation to find the second derivative, using

        solve(%,diff(y(x),x$2));

We then have to substitute for dy/dx using

        subs(diff(y(x),x)=-y(x)/(x+exp(y(x))),%);

and finally

        simplify(%);

In the penultimate command we have typed in the expression for dy/dx, to mirror that substitution step in the algebraic process. In fact it is possible to use MAPLE to avoid having to do this, but we shall not discuss that here.

4.2 Logarithmic Differentiation

This topic is an application of implicit differentiation. It is a technique which is useful when we have expressions involving the variable in an exponent. It can also be applied to complicated products.

Example 4.4

Differentiate $y = x^{\sin x}$.

We take logarithms of both sides of the equation, to give

$$\ln y = \ln \left(x^{\sin x} \right) = \sin x . \ln x.$$

We deal with the left hand side using implicit differentiation, and the right hand side using the product rule. This gives

$$\frac{1}{y} \frac{dy}{dx} = \cos x . \ln x + \sin x . \frac{1}{x}.$$

We therefore deduce that

$$\frac{dy}{dx} = y \left(\cos x . \ln x + \sin x . \frac{1}{x} \right) = x^{\sin x} \left(\cos x . \ln x + \sin x . \frac{1}{x} \right).$$

Example 4.5

Differentiate $y = a^x$.

In fact we have already encountered this function, in Example 3.9, where we used the definition of a^x in terms of the exponential function. It is worth noting again that we cannot use the rule for differentiating powers which applies when the power is a **constant**. Using that rule would give $\frac{dy}{dx} = xa^{x-1}$ and this is WRONG, as is confirmed if we try to apply the rule to the exponential function. This would give the erroneous calculation

$$\frac{d}{dx} e^x = xe^{x-1},$$

which we know to be incorrect.

On this occasion we obtain the result by logarithmic differentiation, which gives

$$y = a^x; \quad \ln y = x \ln a; \quad \frac{1}{y} \frac{dy}{dx} = \ln a; \quad \frac{dy}{dx} = y \ln a = a^x \ln a.$$

Example 4.6

Differentiate $y = x^2 \sin x \cosh x \, e^x$.

We could use the product rule, but taking logarithms converts the expression into a sum, in which we can differentiate each term separately. Taking logarithms gives

$$\begin{aligned}
\ln y &= \ln(x^2) + \ln(\sin x) + \ln(\cosh x) + \ln(e^x) \\
&= 2\ln x + \ln(\sin x) + \ln(\cosh x) + x.
\end{aligned}$$

Differentiating gives
$$\frac{1}{y}\frac{dy}{dx} = \frac{2}{x} + \cot x + \tanh x + 1.$$

Therefore
$$\begin{aligned}
\frac{dy}{dx} &= y\left(\frac{2}{x} + \cot x + \tanh x + 1\right) \\
&= \left(x^2 \sin x \cosh x\, e^x\right)\left(\frac{2}{x} + \cot x + \tanh x + 1\right). \\
&= 2x \sin x \cosh x\, e^x + x^2 \cos x \cosh x\, e^x \\
&\quad + x^2 \sin x \sinh x\, e^x + x^2 \sin x \cosh x\, e^x.
\end{aligned}$$

Example 4.7

Differentiate $y = \left(x^3 e^x\right)^{\sin x}$.

This example combines a product with an exponent. Taking logarithms gives

$$\ln y = \sin x \ln\left(x^3 e^x\right) = \sin x \left(\ln x^3 + \ln e^x\right) = \sin x \left(3\ln x + x\right).$$

Differentiating with respect to x now gives

$$\frac{1}{y}\frac{dy}{dx} = \cos x\,(3\ln x + x) + \sin x\left(\frac{3}{x} + 1\right),$$

and therefore

$$\frac{dy}{dx} = \left(x^3 e^x\right)^{\sin x}\left(\cos x\,(3\ln x + x) + \sin x\left(\frac{3}{x} + 1\right)\right).$$

4.3 Parametric Differentiation

Equations of curves are often given parametrically, for example the ellipse specified by

$$x = a\cos t, \quad y = b\sin t, \quad 0 \le t \le 2\pi.$$

We want to find the gradient $\dfrac{dy}{dx}$, but the parametric equations can only be differentiated with respect to t.

We can approach this in two ways. Firstly we can use the chain rule to give

$$\frac{dy}{dt} = \frac{dy}{dx}\frac{dx}{dt}, \quad \text{so} \quad \frac{dy}{dx} = \frac{dy}{dt}\bigg/\frac{dx}{dt}, \quad \text{provided} \quad \frac{dx}{dt} \ne 0.$$

Secondly we can go back to the limit definition of the derivative. If we want to work out the gradient $\dfrac{dy}{dx}$ at a point specified by $t = k$, we need to calculate the chord slope limit as $t \to k$. We reason as follows.

$$
\begin{aligned}
\frac{dy}{dx} &= \lim_{t \to k} \frac{y(t) - y(k)}{x(t) - x(k)} = \lim_{t \to k} \frac{y(t) - y(k)}{x(t) - x(k)} \frac{t - k}{t - k} \\
&= \lim_{t \to k} \frac{y(t) - y(k)}{t - k} \frac{t - k}{x(t) - x(k)} \\
&= \lim_{t \to k} \frac{y(t) - y(k)}{t - k} \Big/ \lim_{t \to k} \frac{x(t) - x(k)}{t - k} = \frac{dy}{dt} \Big/ \frac{dx}{dt},
\end{aligned}
$$

provided $\dfrac{dx}{dt} \neq 0$.

Example 4.8

We shall use the formula developed above to find the gradient at an arbitrary point t on the ellipse specified by

$$
x = a \cos t, \quad y = b \sin t, \quad 0 \le t \le 2\pi.
$$

The gradient is given by

$$
\frac{dy}{dx} = \frac{dy}{dt} \Big/ \frac{dx}{dt} = \frac{b \cos t}{-a \sin t} = -\frac{b}{a} \cot t.
$$

The calculation is valid provided $\sin t \neq 0$, which excludes the points given by $t = 0, \pm\pi, \pm 2\pi, \ldots$, where the tangent to the ellipse is parallel to the y-axis.

Example 4.9

Find $\dfrac{dy}{dx}$ given that $x = t^2, y = t^3$.

These are the parametric equations of a curve known as a semicubical parabola. Its graph is shown in Figure 11.2, where we calculate the length of part of this curve.

The derivative is given by

$$
\frac{dy}{dx} = \frac{dy}{dt} \Big/ \frac{dx}{dt} = \frac{3t^2}{2t} = \frac{3t}{2} \quad (t \neq 0).
$$

In this case we can eliminate the parameter t to give $y^2 = x^3$, and so we could also find the derivative using implicit differentiation, as follows.

$$
2y \frac{dy}{dx} = 3x^2, \quad \text{so} \quad \frac{dy}{dx} = \frac{3x^2}{2y} \quad (y \neq 0).
$$

We can see that the graph in Figure 11.2 is not the graph of a function, and so we would need values of both x and y to specify a point of the curve, and hence find the gradient. With the parametric form, a given value of t determines both x and y, and hence a unique point on the curve. That value of t will determine the gradient at that point.

Example 4.10

Given $x = a \cos t$, $y = b \sin t$, $0 \leq t \leq 2\pi$, find $\dfrac{d^2y}{dx^2}$.

It is possible to find a general formula for the second derivative, but it is clearer to argue as follows.

We recall that $\dfrac{d^2y}{dx^2} = \dfrac{dY}{dx}$, where $Y = \dfrac{dy}{dx}$.

Applying the parametric differentiation formula to Y gives

$$\frac{dY}{dx} = \frac{dY}{dt} \bigg/ \frac{dx}{dt}.$$

We worked out Y in Example 4.8, and so we apply this formula, giving

$$\frac{d^2y}{dx^2} = \frac{dY}{dx} = \frac{dY}{dt} \bigg/ \frac{dx}{dt} = -\frac{b}{a} \left(\operatorname{cosec}^2 t \right) / (-a \sin t) = -\frac{b}{a^2 \sin^3 t},$$

provided $\sin t \neq 0$.

NOTE: A common mistake is to try to find $\dfrac{d^2y}{dx^2}$ by differentiating the formula obtained for $\dfrac{dy}{dx}$ with respect to t. This is WRONG.

4.4 Differentiating Inverse Functions

Inverse functions were discussed in some detail in Section 1.7, and we now consider their differentiation. It is possible to find a general formula, as we shall demonstrate, but in most cases it is more helpful to use an implicit function approach, and this is done in the examples in this section.

Suppose that we have a differentiable function f with its inverse g. So $y = f(x)$ and $x = g(y)$ are equivalent. We shall establish differentiability of g using the limit definition.

$$\frac{dg}{dy} = \lim_{k \to 0} \frac{g(y+k) - g(y)}{k}.$$

Now $y + k = f(x + h)$ for some h, and since f is continuous it follows that $k \to 0$ as $h \to 0$. Also, since f has an inverse, it is 1-1, so for $h \neq 0$ we have $f(x + h) \neq f(x)$, so that $k \neq 0$. Therefore

$$\frac{g(y + k) - g(y)}{k} = \frac{g(y + k) - g(y)}{y + k - y} = \frac{x + h - x}{f(x + h) - f(x)} = \frac{h}{f(x + h) - f(x)}.$$

From this we deduce that

$$\frac{dg}{dy} = \lim_{k \to 0} \frac{g(y + k) - g(y)}{k} = \lim_{h \to 0} \frac{h}{f(x + h) - f(x)} = 1 \bigg/ \frac{df}{dx}.$$

If we were to assume that the inverse g is differentiable then we could obtain the same formula from the inverse function relationship $g(f(x)) = x$. Differentiating this equation using the chain rule gives $g'(f(x))f'(x) = 1$, and therefore, since $y = f(x)$,

$$g'(y) = \frac{1}{f'(x)}.$$

Example 4.11

We can verify the above rule using the logarithmic function.

Suppose $y = f(x) = \ln x$, so that $x = g(y) = e^y$ is the inverse. then

$$g'(y) = e^y = e^{\ln x} = x = \frac{1}{\frac{1}{x}} = \frac{1}{f'(x)}.$$

Example 4.12

Find the derivative of $\sinh^{-1} x$.

Suppose $y = \sinh^{-1} x$, so that $x = \sinh y$. Differentiating the latter equation implicitly with respect to x gives

$$1 = \cosh y \frac{dy}{dx} \quad \text{so that} \quad \frac{dy}{dx} = \frac{1}{\cosh y},$$

as the general formula above implies. However we want the answer in terms of x, and so we have to find $\cosh y$ in terms of $x = \sinh y$. Using the hyperbolic identity $\cosh^2 y - \sinh^2 y = 1$ gives $\cosh y = \sqrt{1 + \sinh^2 y}$, where we use the positive square root because $\cosh y$ is always positive. Therefore

$$\frac{dy}{dx} = \frac{1}{\cosh y} = \frac{1}{\sqrt{1 + \sinh^2 y}} = \frac{1}{\sqrt{1 + x^2}}.$$

Example 4.13

Let $f(x) = x^3 + 2x - 2$, which is a 1-1 function. Find the derivative of $f^{-1}(x)$ at the point where f and its inverse intersect.

The graphs of the function and its inverse are shown in Figure 4.2.

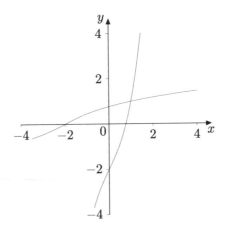

Figure 4.2   Graph of $x^3 + 2x - 2$ and its inverse

The two graphs intersect at the point $(1, 1)$, as can be seen from the fact that $f(1) = 1$. Calculating the derivative gives $f'(x) = 3x^2 + 2$, and so $f'(1) = 5$. Therefore at the point of intersection the derivative of the inverse function has value $1/5$.

Example 4.14

In this example we consider the problem of differentiating the inverse sine function. In Section 1.7.2 we considered the problems involved in restricting the domain of sine so as to obtain a 1-1 function, which would therefore have an inverse. We have to consider the same approach here.

Suppose that $y = \sin^{-1} x$, which is equivalent to $x = \sin y$. Differentiating the latter equation implicitly with respect to x gives

$$1 = \cos y \frac{dy}{dx} \quad \text{so that} \quad \frac{dy}{dx} = \frac{1}{\cos y}.$$

We want the result in terms of x, and so we use the identity $\cos^2 y + \sin^2 y = 1$, giving $\cos y = \pm\sqrt{1 - \sin^2 y} = \pm\sqrt{1 - x^2}$. Unlike the previous example, where sinh is 1-1 over its whole domain, and where the choice of square root was straightforward, in this case we have to consider how the domain is restricted in the same way as in Section 1.7.2. Recall that in that section we used the notation $\sin^{-1} x$ to denote the inverse of the function specified by

$$f(x) = \sin x; \quad -\frac{\pi}{2} \le x \le \frac{\pi}{2}.$$

The graph was shown in Figure 1.22, and we can see that the gradient of the inverse is positive, which means we have to choose the positive square root. We note also that the derivative of sin is cos, which itself is positive in the interval $\left(-\frac{\pi}{2}, \frac{\pi}{2}\right)$, again confirming the choice of the positive square root. So with f specified with the given domain we have

$$\frac{d}{dx}\left(f^{-1}(x)\right) = \frac{1}{\sqrt{1-x^2}}.$$

If on the other hand we consider the function g specified by

$$g(x) = \sin x; \quad \frac{\pi}{2} \le x \le \frac{3\pi}{2},$$

then the gradient of the inverse is negative, as shown in Figure 1.23, and also confirmed by the fact that cosine is negative in the interval $\left(\frac{\pi}{2}, \frac{3\pi}{2}\right)$. So in this case we have

$$\frac{d}{dx}\left(g^{-1}(x)\right) = -\frac{1}{\sqrt{1-x^2}}.$$

4.5 Leibniz Theorem

We saw in Example 3.14 that finding the n-th derivative of a product can be complicated. In this section we derive a general formula for this procedure. If we begin by applying the product rule three times to the general expression of the form $h(x) = f(x)g(x)$ and collect like terms together at each stage we soon perceive a pattern emerging. We find that

$$
\begin{aligned}
h'(x) &= f'(x)g(x) + f(x)g'(x); \\
h''(x) &= [f''(x)g(x) + f'(x)g'(x)] + [f'(x)g'(x) + f(x)g''(x)] \\
&= f''(x)g(x) + 2f'(x)g'(x) + f(x)g''(x); \\
h'''(x) &= [f'''(x)g(x) + f''(x)g'(x)] + 2[f''(x)g'(x) + f'(x)g''(x)] \\
&\quad + [f'(x)g''(x) + f(x)g'''(x)] \\
&= f'''(x)g(x) + 3f''(x)g'(x) + 3f'(x)g''(x) + f(x)g'''(x),
\end{aligned}
$$

where at each stage the square brackets indicate a pair of terms arising from the application of the product rule to a single term at the previous stage.

The pattern of numerical coefficients when the terms are collected together is that of the binomial coefficients from Pascal's triangle, and this enables us to formulate the general result.

Theorem 4.15 (Leibniz)

If the functions $f(x), g(x)$ are both differentiable n times then their product is differentiable n times and

$$
\begin{aligned}
\frac{d^n}{dx^n}(fg) &= \frac{d^n f}{dx^n}g + \binom{n}{1}\frac{d^{n-1}f}{dx^{n-1}}\frac{dg}{dx} + \binom{n}{2}\frac{d^{n-2}f}{dx^{n-2}}\frac{d^2 g}{dx^2} + \cdots \\
&+ \binom{n}{k-1}\frac{d^{n-(k-1)}f}{dx^{n-(k-1)}}\frac{d^{k-1}g}{dx^{k-1}} + \binom{n}{k}\frac{d^{n-k}f}{dx^{n-k}}\frac{d^k g}{dx^k} + \cdots + f\frac{d^n g}{dx^n} \\
&= \sum_{k=0}^{n}\binom{n}{k}\frac{d^{n-k}f}{dx^{n-k}}\frac{d^k g}{dx^k}.
\end{aligned}
$$

Proof

The pattern we established above provides evidence for the truth of the result. We include a proof here for readers who are familiar with proof by induction (see Howie, Chapter 1) and the basic properties of binomial coefficients

In the course of the proof we use relationships involving binomial coefficients, which we prove first. It may be helpful to remind readers of the definition and notation for binomial coefficients. They occur in binomial expansions such as $(1+x)^n$, where n is a positive integer. The k-th binomial coefficient is the coefficient of x^k in this expansion. It is given by the following formula

$$
\binom{n}{k} = \frac{n!}{(n-k)!k!},
$$

for $k = 0, 1, \ldots, n$, where $0!$ is defined to be 1. In fact for $k = 0$ and $k = n$ respectively we have

$$
\binom{n}{0} = \frac{n!}{(n-0)!0!} = 1, \quad \binom{n}{n} = \frac{n!}{(n-n)!n!} = 1.
$$

These expressions occur in Pascal's Triangle, and the following addition rule expresses in general terms the way in which we obtain the coefficients in a particular row by adding the two appropriate entries from the row above.

$$
\binom{n}{k-1} + \binom{n}{k} = \frac{n!}{(n-k+1)!(k-1)!} + \frac{n!}{(n-k)!k!}
$$

$$= \frac{n!}{(n-k+1)!k!}(k+(n-k+1))$$

$$= \frac{(n+1)!}{(n+1-k)!k!} = \binom{n+1}{k}.$$

We note the particular case $k = 1$, which we use below. This states that

$$\binom{n}{0} + \binom{n}{1} = \binom{n+1}{1}, \quad \text{i.e.,} \quad 1 + \binom{n}{1} = \binom{n+1}{1}.$$

The proof of Leibniz Theorem uses the method of mathematical induction. The result is true for $n = 1$ because it is just the ordinary product rule. If the result is true for n as in the statement of the theorem then we differentiate both sides once more with respect to x. Each term on the right hand side gives rise to two terms, from the product rule. We therefore have

$$
\begin{aligned}
\frac{d^{n+1}}{dx^{n+1}}(fg) =\; & \frac{d^{n+1}f}{dx^{n+1}}g + \frac{d^n f}{dx^n}\frac{dg}{dx} \\
& + \binom{n}{1}\frac{d^n f}{dx^n}\frac{dg}{dx} + \binom{n}{1}\frac{d^{n-1}f}{dx^{n-1}}\frac{d^2 g}{dx^2} \\
& + \binom{n}{2}\frac{d^{n-1}f}{dx^{n-1}}\frac{d^2 g}{dx^2} + \binom{n}{2}\frac{d^{n-2}f}{dx^{n-2}}\frac{d^3 f}{dx^3} + \cdots \\
& + \binom{n}{k-1}\frac{d^{n-k+2}f}{dx^{n-k+2}}\frac{d^{k-1}g}{dx^{k-1}} + \binom{n}{k-1}\frac{d^{n-k+1}f}{dx^{n-k+1}}\frac{d^k g}{dx^k} \\
& + \binom{n}{k}\frac{d^{n-k+1}f}{dx^{n-k+1}}\frac{d^k g}{dx^k} + \binom{n}{k}\frac{d^{n-k}f}{dx^{n-k}}\frac{d^{k+1}g}{dx^{k+1}} + \cdots \\
& + \frac{df}{dx}\frac{d^n g}{dx^n} + f\frac{d^{n+1}g}{dx^{n+1}}.
\end{aligned}
$$

We now rearrange the terms in pairs so that they contain the same derivative. So the second term in line 1 of the above chain of expressions combines with the first term in line 2, the second term in line 2 with the first term in line 3, and so on. This now gives

$$
\begin{aligned}
\frac{d^{n+1}}{dx^{n+1}}(fg) =\; & \frac{d^{n+1}f}{dx^{n+1}}g \\
& + \frac{d^n f}{dx^n}\frac{dg}{dx} + \binom{n}{1}\frac{d^n f}{dx^n}\frac{dg}{dx} \\
& + \binom{n}{1}\frac{d^{n-1}f}{dx^{n-1}}\frac{d^2 g}{dx^2} + \binom{n}{2}\frac{d^{n-1}f}{dx^{n-1}}\frac{d^2 g}{dx^2} \\
& + \binom{n}{2}\frac{d^{n-2}f}{dx^{n-2}}\frac{d^3 f}{dx^3} + \cdots
\end{aligned}
$$

$$+ \binom{n}{k-1} \frac{d^{n-k+2}f}{dx^{n-k+2}} \frac{d^{k-1}g}{dx^{k-1}}$$

$$+ \binom{n}{k-1} \frac{d^{n-k+1}f}{dx^{n-k+1}} \frac{d^k g}{dx^k} + \binom{n}{k} \frac{d^{n-k+1}f}{dx^{n-k+1}} \frac{d^k g}{dx^k}$$

$$+ \binom{n}{k} \frac{d^{n-k}f}{dx^{n-k}} \frac{d^{k+1}g}{dx^{k+1}} + \cdots$$

$$+ \frac{df}{dx}\frac{d^n g}{dx^n} + f\frac{d^{n+1}g}{dx^{n+1}}.$$

Finally we utilise the addition rule for binomial coefficients to give

$$\frac{d^{n+1}}{dx^{n+1}}(fg) \quad = \quad \frac{d^{n+1}f}{dx^{n+1}}g + \binom{n+1}{1} \frac{d^{n+1-1}f}{dx^{n+1-1}}\frac{dg}{dx}$$

$$+ \binom{n+1}{2} \frac{d^{n+1-2}f}{dx^{n+1-2}}\frac{d^2 g}{dx^2} + \cdots$$

$$+ \binom{n+1}{k} \frac{d^{n+1-k}f}{dx^{n+1-k}}\frac{d^k g}{dx^k} + \cdots + f\frac{d^{n+1}g}{dx^{n+1}},$$

which is the result for $n+1$, thereby completing the proof by induction.

                                                        □

Example 4.16

Find a formula for the n-th derivative of $x^2 \ln(2x + 3)$.

Let $f(x) = \ln(2x+3)$; $g(x) = x^2$. Notice that for g the third and subsequent derivatives are all zero, so that only the first three terms in Leibniz formula are non-zero. We use the formula for the n-th derivative of $f(x)$ which we obtained in Example 3.13, namely

$$f^{(n)}(x) = (n-1)!\frac{(-1)^{(n+1)}2^n}{(2x+3)^n}.$$

Leibniz Theorem therefore gives

$$(fg)^{(n)}(x) \quad = \quad f^{(n)}(x).x^2 + \binom{n}{1} f^{(n-1)}(x).2x + \binom{n}{2} f^{(n-2)}.2$$

$$= \quad (n-1)!\frac{(-1)^{n+1}2^n}{(2x+3)^n}.x^2$$

$$+ \quad n.(n-2)!\frac{(-1)^n 2^{n-1}}{(2x+3)^{n-1}}.2x$$

$$+ \quad \frac{n(n-1)}{2!}.(n-3)!\frac{(-1)^{n-1}2^{n-2}}{(2x+3)^{n-2}}.2.$$

If we take out the factor of $\dfrac{(n-3)!(-1)^{(n+1)}2^{n-2}}{(2x+3)^n}$ from each of the three terms above we are left with

$$(n-1)(n-2).2^2.x^2 - n(n-2).2.2x(2x+3) + n(n-1)(2x+3)^2,$$

which simplifies to $8x^2 + 12nx + 9n^2 - 9n$. We have therefore shown that

$$\frac{d^n}{dx^n}\left(\ln(2x+3)\right) = \frac{(n-3)!(-1)^{(n+1)}2^{n-2}}{(2x+3)^n}\left(8x^2 + 12nx + 9n^2 - 9n\right).$$

EXERCISES

4.1. For each of the following, find $\dfrac{dy}{dx}$ in terms of x and y.

(a) $2xy + x - 3y = 2$;

(b) $(x+1)^2 + 2(y-1)^3 = 0$;

(c) $x^3y^3 = xy - 1$;

(d) $y\ln x = x\ln y$;

(e) $x^2 - 3xy^2 + y^3x - y^2 = 2$;

(f) $xy\sqrt{x+y} = 1$;

(g) $\dfrac{x}{y} - \dfrac{y}{x} = 1$;

(h) $\dfrac{x+y}{x-y} = \dfrac{x}{y} + \dfrac{1}{y^2}$;

(i) $\sin(xy^2) = x + \cos(yx^2)$;

(j) $xy\exp\left(\dfrac{x}{y}\right) = 1$.

4.2. For each of the following, find $\dfrac{dy}{dx}$ and $\dfrac{d^2y}{dx^2}$ in terms of x and y.

(a) $xy = 2x - 3y$; (b) $x\sin y = \sin x$.

4.3. Given that $x^2 + 2y^2 = 4$, find $\dfrac{d^2y}{dx^2}$ in terms of y only.

4.4. Find the gradient at the point $(2, -2)$ of the curve whose equation is $x^3 - xy - 3y^2 = 0$. Hence determine the equation of the tangent to the curve at that point.

4.5. Find the gradient at the point $(1, 0)$ of the curve whose equation is $x\sin(xy) = x^2 - 1$. Hence determine the equation of the tangent to the curve at that point. Explain from the formula why the curve is symmetric about the y-axis, and hence write down the equation of the tangent at the point $(-1, 0)$.

Verify the symmetry by plotting the graph using MAPLE.

4.6. For each of the following, use logarithmic differentiation to find $\dfrac{dy}{dx}$, expressing the results in terms of x.

(a) $y = x^x$;   (b) $y = x^{-x}$;

(c) $y = (-x)^{-x}$;   (d) $y = (\sin x)^{\sin x}$;

(e) $y = (e^x)^{\ln x}$;   (f) $y = (\ln x)^x$;

(g) $y = (\tan x)^{2x}$;   (h) $y = 2^{x+x^2}$;

(i) $y = \sqrt{e^x \sin x}$;   (j) $y = \sqrt{(x-1)^2 e^{-x} \cos x}$.

4.7. For each of the following, find $\dfrac{dy}{dx}$ and $\dfrac{d^2 y}{dx^2}$ in terms of t. Plot each of the curves using MAPLE.

(a) $x = 1 + \ln t$, $y = t^2 - t$;   (b) $x = t + t^2$, $y = t - t^2$;

(c) $x = t \ln t$, $y = 2t + 3$;   (d) $x = t^2$, $y = e^t + 1$;

(e) $x = t^3$, $y = \sqrt{t^2 + 1}$;   (f) $x = \sin(t^2)$, $y = \cos t$;

(g) $x = \tan t$, $y = e^t$;   (h) $x = \ln(\cos t)$, $y = \sin t$;

(i) $x = \cos t$, $y = te^t$;   (j) $x = \sin(t)$, $y = t^2 + 1$.

4.8. Find an expression for the gradient $\dfrac{dy}{dx}$, in terms of t, at a point on the hyperbola given by

$$x = a \cosh t, \quad y = b \sinh t.$$

Write down any values of t for which the gradient is undefined, explaining the geometrical significance.

Find expressions for $\dfrac{d^2 y}{dx^2}$ and $\dfrac{d^3 y}{dx^3}$, in terms of t.

4.9. Find an expression, in terms of t, for the gradient at a point on the curve specified by

$$x = t \cos t, \quad y = t \sin t, \quad t \geq 0.$$

Plot the curve using MAPLE and explain why there are infinitely many values of t for which the gradient is undefined.

4.10. For each of the following functions, none of which is 1-1, investigate differentiation of inverse functions obtained by restricting the
domain in various ways, as in Example 4.14.

(a) $f(x) = \cosh x$; (b) $f(x) = \tan x$; (c) $f(x) = e^{x^2}$.

4.11. Find an expression for the derivative of $\tanh^{-1} x$.

4.12. Use Leibniz Theorem to find the n-th derivative of each of the following.

(a) $x \ln x$; (b) $(x^2 - 2x + 3)e^{2x}$; (c) $x^3 e^{-x}$.

5
Applications of Differentiation

5.1 Gradients and Tangents

We saw in Section 3.1 that $f'(a)$, the value of the derivative $f'(x)$ at $x = a$, corresponds to the gradient of the tangent to the graph of $y = f(x)$ at the point $(a, f(a))$. This enables us to write down the equation for the tangent line at $(a, f(a))$ in the form

$$y - f(a) = f'(a)(x - a).$$

This is used in many problems in co-ordinate geometry, as in the following examples.

Example 5.1

Let $A = (-a, 0)$ be a point on the negative y-axis (so $a > 0$). Find the equations of the tangents to the parabola $y = x^2$ which pass through the point A. Find the distance between the points where these tangents cross the x-axis.

The equation of the tangent at a point (p, p^2) on the parabola is

$$y - p^2 = 2p(x - p).$$

This can be rearranged as

$$y = 2px - p^2.$$

This meets the y-axis where $x = 0$, namely $y = -p^2$. This should be the point A, so we require $p^2 = a$, i.e., $p = \sqrt{a}$.

The tangent crosses the x-axis where $y = 0$, i.e., where $x = \dfrac{p}{2}$.

Figure 5.1 (which corresponds to the case $a = 4$) shows the symmetry, and so we conclude that the distance between the points where the two tangents cross the x-axis is $p = \sqrt{a}$.

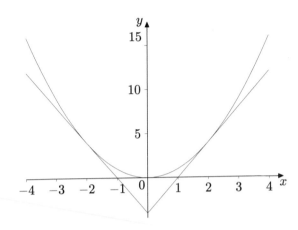

Figure 5.1   Parabola and tangents

Example 5.2

Let A be a point on the ellipse given by $x = a \cos t$, $y = b \sin t$. B is the reflection of A in the y-axis. The tangents at A and B cross at right-angles. Find the values of t.

The points A and B have coordinates $(\pm a \cos t, b \sin t)$, which can be written in the form
$$(a \cos t, b \sin t), \quad (a \cos(\pi - t), b \sin(\pi - t)).$$

To find the gradients we need to use parametric differentiation, and we see from Example 4.8 that the gradients are
$$-\frac{b}{a} \cot t \quad \text{and} \quad -\frac{b}{a} \cot(\pi - t)$$

respectively. For the tangents to cross at right-angles the product of the gradients must be -1. Therefore
$$\left(\frac{b}{a} \cot t \right) \left(\frac{b}{a} \cot(\pi - t) \right) = -1.$$

Using the fact that $\cot(\pi - t) = -\cot t$ we have

$$\frac{b^2}{a^2} \cot^2 t = 1, \quad \text{i.e.,} \quad \tan t = \pm\frac{b}{a},$$

and the values of t required are the solutions of this equation. By way of example, if the ellipse is in fact a circle, so that $a = b$, then $\tan t = \pm 1$, so $t = \pm\frac{\pi}{4}$, as would be clear from a diagram.

5.2 Maxima and Minima

Finding how large or small a given quantity can be is important in many problems in science, engineering, economics, etc., and when this quantity is given by an algebraic formula we can use differentiation to determine these maximum or minimum values. We need to distinguish various kinds of maxima and minima: the following definition does this.

Definition 5.3

Suppose that f is a function with domain D, and that a is a number in D.
   (i) f is said to have a **global maximum** at a if
          $f(x) \leq f(a)$ for all x in D.
   (ii) f is said to have a **global minimum** at a if
          $f(x) \geq f(a)$ for all x in D.
   (iii) f is said to have a **local maximum** at a if
          there is an interval $(a - \delta, a + \delta)$ such that $f(x) \leq f(a)$
          for all x in D belonging to this interval.
   (iv) f is said to have a **local minimum** at a if
          there is an interval $(a - \delta, a + \delta)$ such that $f(x) \geq f(a)$
          for all x in D belonging to this interval.
   We can see from these definitions that a global maximum/minimum is a local maximum/minimum, but not necessarily the converse.
   In some texts "absolute" is used in place of "global" and "relative" is used in place of "local".

Example 5.4

Locate the maxima and minima of the function defined by $f(x) = x^3 - 3x^2 - 9x$, for $-4 \leq x \leq 6$, by plotting its graph.

In this example we have set the domain D to be the interval specified by $-4 \leq x \leq 6$. We can see from the graph that f has a global minimum at $x = -4$. It is the place where the value of $f(x)$ is smallest compared with all other values. Likewise there is a global maximum at $x = 6$, because $f(6)$ is larger than all other values of $f(x)$, where x belongs to D.

When $x = -1$ we have a local maximum. We can see from the graph that $f(x) \leq f(-1)$ for all x satisfying $-2 \leq x \leq 0$. This is not a global maximum because, for example, $f(6) > f(-1)$.

When $x = 3$ we have a local minimum. We can see from the graph that $f(x) \geq f(3)$ for all x satisfying $0 \leq x \leq 4$. This is not a global minimum because for example $f(-4) < f(3)$.

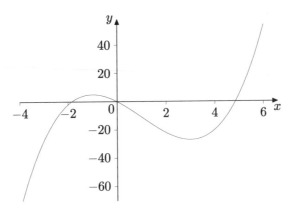

Figure 5.2   Maximum and minimum

To complete the description of the maxima and minima we have to calculate the value of the function at each of the corresponding values of x.

So to summarise:

There is a local maximum at the point $(-1, f(-1)) = (-1, 5)$.

There is a local minimum at the point $(3, f(3)) = (3, -27)$.

There is a global maximum at the point $(6, f(6)) = (6, 54)$.

There is a global minimum at the point $(-4, f(-4)) = (-4, -76)$.

We shall now see how to find such points using differentiation.

Theorem 5.5

Suppose that $f(x)$ is differentiable at $x = a$ and that $x = a$ is a local maximum or a local minimum. Then $f'(a) = 0$.

Proof

We use the limit definition of the derivative (Definition 3.1), namely

$$f'(a) = \lim_{x \to a} \frac{f(x) - f(a)}{x - a}.$$

This is a two-sided limit, and so both one-sided limits will be the same. So

$$f'(a) = \lim_{x \to a^+} \frac{f(x) - f(a)}{x - a} = \lim_{x \to a^-} \frac{f(x) - f(a)}{x - a}.$$

Suppose there is a local maximum at $x = a$. Then $f(x) \leq f(a)$ for all x in some interval I containing a. So for all x in this interval we have $f(x) - f(a) \leq 0$. For the right-sided limit we have $x - a > 0$ and so for x in I we deduce that $\frac{f(x) - f(a)}{x - a} \leq 0$. We are considering the limit as $x \to a$, and so x can be restricted to the interval I. Therefore

$$f'(a) = \lim_{x \to a^+} \frac{f(x) - f(a)}{x - a} \leq 0.$$

For the left-sided limit we have $x - a < 0$ and so for x in I we deduce that $\frac{f(x) - f(a)}{x - a} \geq 0$. Therefore

$$f'(a) = \lim_{x \to a^-} \frac{f(x) - f(a)}{x - a} \geq 0.$$

We therefore conclude that $f'(a) = 0$.

The case of a local minimum is proved in exactly the same way.   □

The converse of this theorem is not true, and there are situations where the derivative is zero but we do not have a maximum or a minimum. One of the simplest examples is $f(x) = x^3$. Differentiating gives $f'(x) = 3x^2$, which is zero when $x = 0$. This point is not a maximum or a minimum, because x^3 is negative for all $x < 0$ and positive for all $x > 0$. Nevertheless, points such as this are also important, in interpreting graphs for example, and we group them all together in the following definition.

Definition 5.6

Given a differentiable function $f(x)$, any point a for which $f'(a) = 0$ is called a **stationary point**. Other terms sometimes used are **turning point** and **critical point**.

Once we have found all the stationary points by solving $f'(x) = 0$ we need to be able to decide what type they are. We will discuss three possible approaches to this.

Firstly we can simply plot the graph, perhaps using MAPLE. We still need to solve $f'(x) = 0$ because the graph will not tell us the exact solutions, which may involve irrational numbers such as $\sqrt{3}$ or π. Once we know where the stationary points are then the graph should be able to tell us whether they are maxima, minima or neither. Naturally we need to choose carefully the values of x and y to be included in the plot, so that we obtain as clear a picture as we can of the behaviour near to the stationary point under investigation.

The second method involves consideration of the derivative near to the stationary point. Near a local maximum, the graph is increasing to the left of the stationary point, and decreasing to the right, so we should expect the derivative to be positive on the left and negative on the right. The reverse occurs with a local minimum.

In the example $f(x) = x^3$ we have $f'(x) = 3x^2$, which is positive both on the left and the right of $x = 0$, and the stationary point is neither a maximum nor a minimum. Such stationary points are sometimes called **points of inflection**.

The third method involves using the second derivative.

Theorem 5.7 (The second derivative test)

Suppose that $f(x)$ has a stationary point at $x = a$ and that f is twice differentiable at $x = a$.

(i) If $f''(a) > 0$ then $x = a$ is a local minimum.

(ii) If $f''(a) < 0$ then $x = a$ is a local maximum.

(iii) If $f''(a) = 0$ this gives no information about the type of stationary point.

Proof

We use the limit definition

$$f''(a) = \lim_{x \to a} \frac{f'(x) - f'(a)}{x - a}.$$

In this case we have $f'(a) = 0$ and so

$$f''(a) = \lim_{x \to a} \frac{f'(x)}{x - a}.$$

(i) If $f''(a) > 0$ this tells us that for x near to a we must have $f'(x) < 0$ if $x < a$ and $f'(x) > 0$ if $x > a$. The second method described above tells us that we must have a minimum at $x = a$.

(ii) If $f''(a) < 0$ then similar reasoning to (i) tells us that we have a maximum at $x = a$.

(iii) Here we give examples. We have already seen that if $f(x) = x^3$ we have $f''(0) = 0$ but zero is neither a maximum nor a minimum. If $f(x) = x^4$ then clearly we have a minimum at $x = 0$, but $f''(x) = 12x^2$, which is zero at $x = 0$. Likewise if $f(x) = -x^4$ then clearly we have a maximum at $x = 0$, but $f''(x) = -12x^2$, which again is zero at $x = 0$. This example demonstrates that the condition $f''(a) = 0$ is inconclusive.

$\square$

We note that it is normally possible to investigate case (iii) using Taylor series, discussed in Chapter 6.

Example 5.8

Locate and classify maxima and minima for the function defined by

$$f(x) = x^5 - 4x^3, \quad (-2.2 \le x \le 2.2).$$

We need to find where $f'(x) = 0$, so we have

$$f'(x) = 5x^4 - 12x^2 = x^2 \left(5x^2 - 12\right) = 0 \quad \text{when} \quad x = 0, \pm\sqrt{12/5}.$$

We shall try to classify these stationary points using the second derivative test. We have $f''(x) = 20x^3 - 24x = x(20x^2 - 24)$. Evaluating this at the stationary points gives

$f''\left(\sqrt{12/5}\right) = 24\sqrt{12/5} > 0$ so we have a local minimum at $x = \sqrt{12/5}$.

$f''\left(-\sqrt{12/5}\right) = -24\sqrt{12/5} < 0$ and therefore we have a local maximum at $x = -\sqrt{12/5}$.

$f''(0) = 0$ so the second derivative test gives no information here. To investigate this further we can plot the graph.

We can see from Figure 5.3 that the local maximum and local minimum we found are confirmed, and that the stationary point at $x = 0$ is neither a maximum nor a minimum. In fact since $f'(x) = x^2 \left(5x^2 - 12\right)$ we can see that the derivative is negative near to $x = 0$ on both sides, so that the graph is decreasing in an interval around $x = 0$.

The other thing we can see from the graph is that there is a global minimum at the left hand endpoint of the domain, and a global maximum at the right hand endpoint of the domain.

To verify this numerically we need to find the values of the function at the stationary points and at the endpoints. We find that

$$f(2.2) = (2.2)^5 - 4(2.2)^3 \approx 8.94,$$

$$f\left(\sqrt{\tfrac{12}{5}}\right) = \left(\sqrt{\tfrac{12}{5}}\right)^5 - 4\left(\sqrt{\tfrac{12}{5}}\right)^3 = -\tfrac{96}{25}\left(\sqrt{\tfrac{12}{5}}\right) \approx -5.95.$$

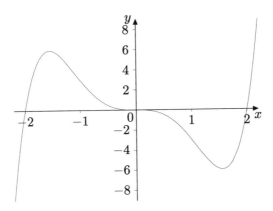

Figure 5.3   Graph of $f(x) = x^5 - 4x^3$

Since f is an odd function we can write down the corresponding values
$f(-2.2) \approx -8.94$,
$f\left(-\sqrt{\frac{12}{5}}\right) = \frac{96}{25}\left(\sqrt{\frac{12}{5}}\right) \approx 5.95$.
This confirms that the global maximum and the global minimum are at the right- and left-hand endpoints of the domain respectively.

In fact whenever we have a continuous function defined on a closed interval, it will always have a global maximum and a global minimum. This result is discussed in Howie, Chapter 3. Global maximum and minimum values may occur at the endpoints, as in this case, or at a stationary point if the value of the function there is greater.

5.3 Optimisation Problems

In this section we are concerned with optimisation problems. This means we are trying to find a solution to a problem which optimises (finds the best value for) some quantity such as volume, or cost. We shall consider problems where the optimum solution corresponds to a maximum or a minimum. Problems such as this are often stated verbally, and we can use a common strategy for their solution, with the following steps.

1. Choose variables to represent the quantities described in the problem.

2. If possible draw a diagram and indicate these quantities on the diagram.

3. Use the diagram and the information given in the problem to find formulae relating the quantities mentioned.

4. Use these formulae to determine a formula for the quantity to be optimised as a function of one variable.

5. Determine any restrictions on the variables arising from the conditions of the problem (for example an area should be positive).

6. Use the techniques of calculus to determine the stationary points, paying attention to any restrictions on the variables.

7. Determine maxima and minima and choose the one which provides the solution to the problem.

In this section we restrict attention to problems involving one variable, although in practice optimisation problems often involve many variables, and their investigation forms part of Operational Research.

Example 5.9

Find the point on the curve $y = x^3$ which is nearest to $(4, 0)$.

The variables here will be the coordinates of an arbitrary point (x, y) on the curve, and the distance D of that point from $(4, 0)$. The variables x and y are of course related through the equation of the curve, $y = x^3$.

We know that the distance will involve a square root, from Pythagoras' Theorem. It is therefore likely to be easier algebraically to consider the square of the distance, so we have $D^2 = (x - 4)^2 + y^2$. Using the relationship $y = x^3$ then gives

$$D^2 = (x - 4)^2 + x^6.$$

We have now obtained a formula for the quantity to be optimised as a function of one variable. To find its minimum value we look for stationary points. Differentiating gives

$$\frac{d(D^2)}{dx} = 2(x - 4) + 6x^5 = 2(x - 1)(3x^4 + 3x^3 + 3x^2 + 3x + 4).$$

The polynomial of degree four is clearly positive for all positive values of x, and it is clear from drawing a diagram that the nearest point to $(4, 0)$ on the curve will be in the first quadrant. So $x = 1$ gives the stationary point we need, and geometrically it is clear that this will be a minimum. The smallest distance is therefore given by $D^2 = (1 - 4)^2 + 1^6 = 10$, so $D = \sqrt{10}$ is the solution to the problem.

The reasoning used here is typical of this kind of problem, being a mixture of calculation and geometrical reasoning derived from the conditions of the problem.

Example 5.10

A rectangular piece of plastic sheeting 10m long and 2m wide is folded in half lengthways to form a tunnel in the shape of a triangular prism to protect plants in the garden from cold weather. What should be the height of the prism to maximise the volume of the prism?

The variables in this problem are the volume V, the height h and half the base b, as shown in Figure 5.4, from which we obtain the following relationships. From Pythagoras' Theorem, $h^2 + b^2 = 1$, and the volume is $V = 10hb$. It is clear from these equations that it will be easier to consider the square of the volume, and we eliminate b to obtain

$$V^2 = 100h^2b^2 = 100h^2(1 - h^2) = 100(h^2 - h^4).$$

Differentiating with respect to h then gives

$$\frac{d(V^2)}{dh} = 100(2h - 4h^3) = 200h(1 - 2h^2) = 200h(1 - h\sqrt{2})(1 + h\sqrt{2}).$$

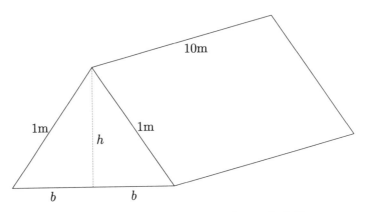

Figure 5.4   Diagram for Example 5.10

We want the derivative to be zero, and from the conditions of the problem we must have h positive, and the only positive value of h giving a zero derivative is $h = \dfrac{1}{\sqrt{2}}$. Now because of the factor $1 - 2h^2$ the derivative changes from positive to negative as h increases through the stationary point. We therefore have a maximum. The maximum volume is given by

$$V = \sqrt{100\left(\frac{1}{2} - \frac{1}{4}\right)} = 5\text{m}^3.$$

Example 5.11

A pharmaceutical company manufactures a vaccine which it sells to hospitals for £20 per dose. The cost of production for x doses is £$(1000 + 5x + 0.002x^2)$. The factory can produce at most 5000 doses in a week. How many doses should the company manufacture and sell in a week to maximise the profit?

The variables in this problem are the profit P, the cost of production C and the sales receipts S. These are related by the simple equation $P = S - C$. Using the information in the problem therefore gives the profit made by selling x doses as

$$P = 20x - (1000 + 5x + 0.002x^2) = -0.002x^2 + 15x - 1000.$$

We must note that of course x is an integer variable. In order to apply the methods of calculus we would need to treat it as a real variable. We would have to be careful if the stationary point turned out not to occur at an integer value. Now in fact, because P is given by a quadratic function, we do not need to use differentiation, we can simply complete the square, giving

$$P = -0.002x^2 + 15x - 1000 = -0.002(x - 3750)^2 + 27125.$$

We can see from this that the stationary point occurs where $x = 3750$, i.e., where $(x - 3750)^2 = 0$, and the maximum profit is £27,125. So the company should manufacture and sell 3750 doses per week to maximise profit. Naturally differentiation would give the same result.

5.4 The Newton-Raphson Method

In this section we consider a numerical method, using the derivative, for solving equations for which we cannot find exact solutions algebraically.

Geometrically the method can be understood through Figure 5.5. We are trying to find the value of x for which the curve $y = f(x)$ crosses the x-axis. In other words we are trying to solve the equation $f(x) = 0$. The method proceeds by having an initial guess x_0, possibly obtained from a graph, which we hope is close to the actual solution. We then join the point $(x_0, 0)$, on the x-axis, by means of a line parallel to the y-axis, to the point $(x_0, f(x_0))$ lying on the curve. We then draw the tangent to the curve at this point, and we denote by x_1 the value of x for which the tangent meets the x-axis. We can see on the diagram that this new value appears to be closer to the actual solution than the first guess x_0. We then continue this procedure, generating a sequence of

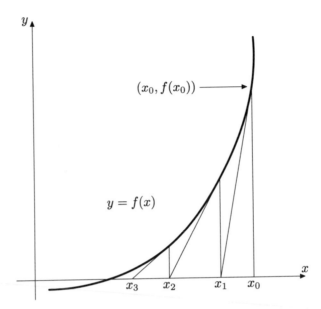

Figure 5.5   The Newton-Raphson method

numbers $x_0, x_1, x_2, \cdots, x_n, \cdots$ which we hope will converge to the solution we are looking for.

To derive a formula which gives a number in the sequence in terms of its predecessor, we note that the gradient of the first tangent we drew is $f'(x_0)$, and so its equation is

$$y - f(x_0) = f'(x_0)(x - x_0).$$

This crosses the x-axis where $y = 0$, so that $-f(x_0) = f'(x_0)(x - x_0)$ which we can rearrange in the form

$$x = x_0 - \frac{f(x_0)}{f'(x_0)}.$$

This only makes sense provided $f'(x_0) \neq 0$, and we can see that if the derivative were zero then the tangent would be parallel to the x-axis and would not intersect it at all. In fact if the derivative is non-zero but very small then the method can break down, firstly because the tangent may intersect the x-axis a very long way away from the solution we are seeking, and secondly because dividing by a very small number on a computer or a calculator can give rise to significant numerical inaccuracy.

Apart from considerations such as these we have now determined the value of x_1. We generalise this to give the iterative formula

$$x_{n+1} = x_n - \frac{f(x_n)}{f'(x_n)}.$$

For a given function $f(x)$ we can find its derivative and obtain a specific Newton-Raphson formula, which lends itself to the use of a spreadsheet for finding successive terms of the sequence $x_0, x_1, x_2, \cdots, x_n, \cdots$, thereby getting an idea of whether the sequence converges, and obtaining an approximate solution to $f(x) = 0$ to as many decimal places as the spreadsheet can work with. It is possible to investigate this further to find bounds for the error at each stage of the approximation, but this is beyond the scope of this book.

Example 5.12

Use the Newton-Raphson method to find the least positive solution of the equation $x = 2 \sin x$.

We need the equation in the form $f(x) = 0$, so we can write

$$f(x) = x - 2 \sin x.$$

We then have $f'(x) = 1 - 2 \cos x$, and so the Newton-Raphson formula gives

$$x_{n+1} = x_n - \frac{x_n - \sin(x_n)}{1 - \cos(x_n)}.$$

Plotting a graph of $f(x)$ gives an approximate answer of 1.9, which we shall use for x_0.

The first few rows of the output from a spreadsheet are shown in the following table, formatted to present 10 decimal places.

n	x_n	x_{n+1}
0	1.9000000000	1.8944080080
1	1.8944080080	1.8957571980
2	1.8957571980	1.8954306858
3	1.8954306858	1.8955096457
4	1.8955096457	1.8954905476
5	1.8954905476	1.8954951666
6	1.8954951666	1.8954940495

This suggests that the solution is 1.89549 correct to 5 decimal places.

5.5 Motion in a Straight Line

Speed records are measured by a vehicle travelling a specified distance and the time taken being recorded. Calculating distance divided by time then gives the average speed over that distance, and this is what the records comprise. However if you drive a car you see a continuous readout on the speedometer, suggesting some idea of speed at an instant of time.

Suppose that we have a car travelling in a straight line, and that $x(t)$ denotes its distance from some fixed point at time t. Suppose also that we can record the distance travelled at two times very close to one another, say t and $t+h$, and calculate the average speed over that short time interval. The average speed will be measured by distance travelled divided by time taken, i.e.,

$$\frac{x(t+h) - x(t)}{h}.$$

This is the quotient whose limit as h tends to 0 is the derivative (the rate of change of distance with respect to time), and so this will correspond to the idea of speed at an instant of time. Now if the vehicle is travelling in the negative direction relative to the measurement of distance, then the derivative would be negative, but the speedometer would still record some positive value like 80km/h. We therefore need to take the direction of motion into account if we wish to use the derivative to describe how fast something is moving. From the discussion above it seems that a speedometer measures the magnitude of the derivative, and we use the word **speed** in this sense. When we want to include the direction of motion we use the word **velocity**.

When we begin a car journey our speed increases. We say that we are accelerating. When we slow down we say that we are decelerating. We describe acceleration mathematically as the rate at which the velocity changes. In this sense, when the car is decelerating its acceleration is negative.

Definition 5.13

If an object is moving along the x-axis so that its distance from the origin at time t is $x(t)$, we define its **velocity** at time t to be $\frac{dx}{dt}$. We define its **speed** at time t to be $\left|\frac{dx}{dt}\right|$. We define the **acceleration** of the object to be $\frac{d^2x}{dt^2}$.

Example 5.14

An object is oscillating up and down while suspended from a spring. Its distance, measured downwards in centimetres below the suspension point P of the spring, is given by $x(t) = 2 + \cos 2t$. Describe the subsequent motion.

We know that cosine varies between ± 1 and so the distance of the object below P varies between 1cm and 3cm. The formulae for velocity and acceleration respectively are given by

$$\frac{dx}{dt} = -2\sin 2t; \quad \frac{d^2x}{dt^2} = -4\cos 2t.$$

All this information can be seen by plotting and interpreting graphs. Figure 5.6 plots the distance of the object below the point of suspension as a function of time. Figure 5.7 shows the velocity as a function of time. The verbal description of the motion given below corresponds to the important features of these graphs.

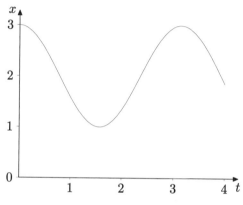

Figure 5.6   Distance-time graph

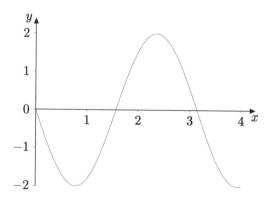

Figure 5.7   Velocity-time graph

Clearly the trigonometric functions correspond to periodic oscillations. When $t = 0$ the object is at its maximum distance below P. The velocity is then zero and the acceleration is negative, with magnitude 4, the maximum possible. This means that at that instant the object has stopped moving. Its speed has been decreasing, and it is about to change direction and pick up speed again, so a deceleration has been taking place, and the negative sign for the second derivative at $t = 0$ confirms that. When $t = \dfrac{\pi}{4}$ the object is at the mid-point of its path, at distance 2cm below P. At that point we have $\sin 2t = 1$ and so the speed has its maximum value. The velocity is negative, meaning that the object is moving upwards as fast as it can at that point. The acceleration is zero there. The object has been speeding up and it is just about to start slowing down.

Example 5.15

A ship leaves port and sails away in a straight line. A radar station is located 3km away from the port in a direction perpendicular to the direction of the ship. When the ship is 5km from the radar station its distance from the radar station is increasing at a rate of 2km/h. What is its velocity at that point?

Let $x(t)$ denote the distance of the ship from port at time t and let $y(t)$ denote its distance from the radar station. Pythagoras' Theorem tells us that $x(t)^2 + 3^2 = y(t)^2$. This is an implicit equation, so we use implicit differentiation to find a relationship between the rates of change. We have

$$2x\frac{dx}{dt} = 2y\frac{dy}{dt} \quad \text{and so} \quad \frac{dx}{dt} = \frac{y}{x}\frac{dy}{dt} \quad (x \neq 0).$$

At the point in question we are given that $y = 5$ and $\dfrac{dy}{dt} = 2$. We therefore have $x = 4$, and so

$$\frac{dx}{dt} = \frac{y}{x}\frac{dy}{dt} = \frac{5}{4} \times 2 = \frac{5}{2}\text{km/h}.$$

Example 5.16

Derive formulae for the motion of an object in a straight line when it is experiencing constant acceleration (such as that due to gravity).

Since the acceleration is constant we have $\dfrac{d^2x}{dt^2} = a$, where a is constant. Basic integration then tells us that $\dfrac{dx}{dt} = at + b$ for some constant b. To determine a value for b we need to know the velocity at some particular point in time. Suppose that the velocity is v_0 when $t = 0$. Therefore $b = v_0$ and so $\dfrac{dx}{dt} = at + v_0$. Integrating again gives $x(t) = \frac{1}{2}at^2 + v_0t + c$ for some constant c. Now suppose that $x = x_0$ when $t = 0$. This tells us that $c = x_0$.

Therefore $x = \frac{1}{2}at^2 + v_0 t + x_0$. The following three formulae are therefore the standard equations used to analyse problems about motion in a straight line under constant acceleration.

$$\frac{d^2 x}{dt^2} = a; \quad \frac{dx}{dt} = at + v_0; \quad x = \frac{1}{2}at^2 + v_0 t + x_0.$$

5.6 Growth and Decay

We sometimes see images of colonies of bacteria in medical programmes on television. The size of the colony seems to grow slowly at first, but the rate of growth gets more and more rapid. The experts tell us that the bacteria divide to produce new bacteria. So for example if there are 10^6 bacteria then after one division there will be twice as many, namely 2.10^6. After another division there will again be twice as many, i.e., 4.10^6. Following this pattern, after n divisions there will be $2^n.10^6$ bacteria (assuming that none die). We can see that the rate of growth depends on how many bacteria there are at any given stage of the process. The number of bacteria in a colony is large, and division happens very quickly. So we model the process by taking the number of bacteria to be $X(t)$, a quantity changing continuously with time, t. With this assumption $X(t)$ is not restricted to integer values, but it means that we can use calculus to model the growth of the colony. It is found that such models correspond closely with such growth processes in nature. The rate of growth of $X(t)$ will be modelled by the derivative, and the assumption that this is proportional to the quantity present at time t leads to the equation

$$\frac{dX}{dt} = kX,$$

where k is the constant of proportionality. This is known as the exponential growth model. Solving such equations is the province of the study of differential equations, which is outside the scope of this book. However in this case we are looking for a function $X(t)$ whose derivative is itself, multiplied by a constant k, and our experience with differentiation tells us that

$$X(t) = Ae^{kt}$$

is the function we are looking for. Now $X(0) = A$, so the constant A represents the size of the colony of bacteria when we start observing at time $t = 0$.

   In a growth process such as this we have $k > 0$, and $Ae^{kt} \to \infty$ as $t \to \infty$. It is clear therefore that the model only applies for a restricted interval of time, for otherwise the colony would eventually weigh more than the earth! What happens of course is that factors come into play which we did not build into the model. Bacteria die, and growth slows as food runs out.

What we need to know for a particular growth process is the value of the constants A and k. These can be determined from measurements, as in the following examples.

Example 5.17

A colony of bacteria weighs 1μg. After 10 seconds it weighs 1.5μg. Determine the rate of growth, assuming an exponential model, and hence deduce the weight of the colony after 20 seconds.

Let $X(t)$ denote the weight of the colony at time t, so $X(t) = Ae^{kt}$. When $t = 0$, $X(0) = A = 1$. When $t = 10$ we know that $X(10) = e^{10k} = 1.5$. Therefore $k = \ln(1.5)/10$. After 20 seconds we have

$$X(t) = e^{20 \cdot \frac{\ln 1.5}{10}} = e^{\ln(1.5^2)} = 1.5^2 = 2.25\mu g.$$

Example 5.18

A large circular water lily leaf floating on the surface of a pond grows, and spreads out in such a way that the larger it becomes, the faster it grows. The gardener measures the area of the leaf, and on returning 20 days later finds that it has quadrupled in size. Assuming exponential growth, how long did it take from the gardener's initial measurement to double in size?

Let $X(t)$ denote the area of the leaf at time t, so $X(t) = Ae^{kt}$. We have $X(0) = A$, and $X(20) = 4A = Ae^{20k}$, so $e^{20k} = 4$, giving $k = \ln(4)/20$. We want to find the value of t for which $X(t) = 2A$, i.e., $2A = Ae^{kt}$. Therefore

$$t = \frac{\ln 2}{k} = \frac{20 \ln 2}{\ln 4} = \frac{20 \ln 2}{2 \ln 2} = 10 \text{ days.}$$

In fact common sense suggests that if it doubled in 10 days it would be quite likely to double again in another 10 days. This tells us that the exponential growth model agrees with our intuitive ideas about such growth processes.

The opposite of growth is decay, and one of the best-known decay processes is the decomposition of radioactive material. We model this situation by assuming that the more atoms there are in a lump of material the more decay decompositions there will be, suggesting that the rate of decay should be proportional to the number of atoms there are in a lump of such material. Again we can write $\dfrac{dX}{dt} = kX$, and because this is decay rather than growth $X(t)$ should decrease as t increases, giving a negative derivative, and therefore a negative value for k. For this reason we often make the negative sign explicit, and write the equation in the form

$$\frac{dX}{dt} = -kX, \quad \text{giving} \quad X(t) = Ae^{-kt}.$$

We speak of exponential decay. The rate of decay of a radioactive material is usually specified by giving its half-life, which is the length of time taken for half the amount of material to decompose. Of course a lump of such material will not actually halve in size, because the radioactive element decays into something else, like lead for example. Some of the transuranic elements are rather evanescent, having a very short half-life. Others have a very long half-life, and unfortunately some of the radioactive waste products from nuclear fission fall into this category.

Example 5.19

A radioactive element has a half-life of H years. Every year a nuclear reactor produces a quantity of this element as waste. How long will it be before this batch of waste contains only 10% of the original amount of the radioactive material?

Let $X(t)$ denote the amount of material remaining after time t. Assuming exponential decay, we have $X(t) = Ae^{-kt}$, where the original amount of material is A. When $t = H$, the half-life, $X(H) = \frac{1}{2}A$. Therefore

$$\frac{A}{2} = Ae^{-kH}, \quad \text{giving} \quad k = \frac{\ln 2}{H}.$$

When $X(t) = X(0)/10$ we have

$$\frac{A}{10} = Ae^{-kt}, \quad \text{so} \quad t = H\frac{\ln 10}{\ln 2} \approx 3.32H.$$

One of the waste products from nuclear reactors is Plutonium 239. This has a half-life of 24,400 years, so the time taken for 90% of a batch of such material to decay will be approximately 81,000 years.

Example 5.20

Newton's law of cooling states that a body cools at a rate which is proportional to the difference between its temperature and the temperature of its surroundings. On a warm day, when room temperature is 24°C, a cup of coffee is poured out at 80°C. After 5 minutes it has cooled to 60°C. How long must I wait if I like my coffee to be at 45°C before I drink it?

Let $T(t)$ denote the temperature of the coffee at time t. Then Newton's law of cooling tells us that $\dfrac{dT}{dt} = -k(T - 24)$. We first note that $\dfrac{dT}{dt} = \dfrac{d}{dt}(T - 24)$, so we can write the equation as

$$\frac{d}{dt}(T - 24) = -k(T - 24), \quad \text{so} \quad T - 24 = Ae^{-kt}.$$

When $t = 0, T = 80$, so $80 = 24 + A$, giving $A = 56$. After 5 minutes, when $T = 60$, the equation tells us that

$$60 - 24 = 36 = 56e^{-5k}, \quad \text{from which} \quad k = \frac{1}{5}\ln\left(\frac{56}{36}\right) = \frac{\ln 56 - \ln 36}{5}.$$

We want to find the value of t for which $T = 45$, and so we require

$$45 - 24 = 21 = 56e^{-kt}, \quad \text{giving} \quad t = 5\,\frac{\ln 56 - \ln 21}{\ln 56 - \ln 36} \approx 11 \text{ minutes.}$$

EXERCISES

5.1. The tangent to the graph of $y = x^3$ at a point P meets the curve again at another point Q. Find the coordinates of Q in terms of the coordinates of P.

5.2. Two circles of radius a are tangent to one another. The two tangents from the centre of one circle touch the other circle at the two points A and B. Find the distance between A and B.

5.3. Locate and classify the stationary points for the functions defined by the following expressions.

(a) $x^2 - 6x - 4$;

(b) $(x^2 - 1)^2$;

(c) $x^4 + 4x^3 - 2x^2 - 12x - 5$;

(d) $\dfrac{x+1}{x^2 + 2}$;

(e) $\dfrac{2x^2 - 3}{x^2 + 1}$;

(f) $x^2 \exp(-x^2)$;

(g) $\ln(1 + x + x^2)$;

(h) $x^{4/3} - 2x^{2/3}$;

(i) $|x^2 - 4|$;

(j) $\dfrac{x^3}{\sqrt{1 + x^6}}$.

5.4. Find the local and global maxima and minima for the functions defined by the following expressions, with the domains indicated.

(a) $x^2 + 2x - 3$, $-2 \le x \le 2$;

(b) $x(x + 1)^2$, $-2 \le x \le 2$;

(c) $\dfrac{2x + 1}{x^2 + 2}$, $-3 \le x \le 3$;

(d) $x\sqrt{3 - x^2}$, $-\sqrt{3} \le x \le \sqrt{3}$;

(e) $x - 2\sin x$, $-\pi \le x \le \pi$;

(f) $\cos 2x + 2\sin x$, $-\pi \le x \le \pi$;

(g) $\dfrac{\cos x}{2 + \sin x}$, $-\frac{\pi}{2} \le x \le \frac{\pi}{2}$;

(h) $x3^{-x}$, $0 \le x \le 1$;

(i) $\dfrac{\ln x}{x^2}$, $1 \le x \le 5$;

(j) $\dfrac{1}{\cosh(x - 1)}$, $-3 \le x \le 3$.

5.5. Locate all the stationary points for the function $f(x) = \cos(e^x)$. How many of these are in the interval $0 \le x \le 10$?

5.6. A hyperbola has equation $x^2 - 2y^2 = 1$. Find the points on this hyperbola which are closest to the point $(0, a)$ on the y-axis.

5.7. Many optimisation problems are formulated in terms of manufacturing boxes, as in this exercise.

A box without a lid is to be made from a 2m×1m rectangular sheet of card by cutting out four equal squares, one from each corner. The resulting rectangles formed at the edges of the card are then folded to make a rectangular box. Find the maximum volume which can be obtained.

5.8. A right circular cone has a total surface area of 4m². Find the dimensions of the cone which give the greatest volume.

5.9. Stainless steel grain silos are to be manufactured, each with a volume of 2000m³. They will rest on a concrete base and consist of a cylindrical wall and a hemispherical top. A stainless steel hemisphere costs three times as much per unit area to manufacture as a cylinder. Find the radius of the cylinder which will minimise the cost of manufacture.

5.10. A high-speed train is making a long journey, of which 1000km is at a constant speed $v$km/h. The cost of fuel per hour at this speed is given by

$$C = 2048 + v^{\frac{3}{2}}.$$

Find the speed at which the train should travel to minimise the cost of this part of its journey.

5.11. Use the Newton-Raphson method to find approximations for the solutions of the equation

$$x^5 - 4x^3 = 2.$$

Plot a graph to help to choose suitable initial approximations corresponding to each of the roots, and use a spreadsheet to obtain 10 decimal places of accuracy.

5.12. A rocket is launched vertically upwards with an initial speed of 1000m/s. It decelerates due to gravity. A tracking station is 10km away from the launch pad. What is the rate of change of the distance of the rocket from the tracking station when it has reached a height of 10km? Give your answer in m/s correct to one decimal place. [Take g, the acceleration due to gravity, as 9.8m/s^2.]

5.13. Radium 226 has a half-life of 1622 years. It decays and changes into Lead 210 (through intermediate elements of relatively short half lives). The radiation from radium is toxic, so it has to be shielded from those nearby. If we have one gram of radium in the laboratory, how long would it take before there was only 1mg (one thousandth of a gram) of radium left in the sample?

5.14. A Chemistry student takes some molten metal from a furnace, but forgets to measure its temperature. Twenty minutes later he remembers, and finds that the temperature of the sample is 250°C. After another 10 minutes its temperature is 150°C. The ambient temperature in the laboratory is 20°C. Assuming Newton's law of cooling, what was the temperature of the metal when it was taken from the furnace?

6

Maclaurin and Taylor Expansions

How does your calculator work out values of the trigonometric functions, or any of the other functions on the various buttons? The circuitry can do addition, and therefore multiplication, which is in essence repeated addition. It can also do subtraction, and division, which can be thought of as repeated subtraction. It can therefore work out values of polynomials like

$$3 - x + 2x^2 + 4x^3 - 2.7x^4$$

for numerical values of x. We therefore need to be able to find polynomials which approximate to the standard functions, and these could then be programmed into the circuits in a calculator. Clearly such an exercise should include an analysis of the greatest possible errors arising with such approximations if the digits in the display are to be accurate. We start with a simple case.

6.1 Linear Approximation

In this section we concentrate on finding linear approximations to a given function over a specified interval.

The curve shown in Figure 6.1 has equation $y = 10 - x^3$, and we have drawn in three of many possible lines which might be considered as providing linear approximations for this function in the interval $0 \leq x \leq 2$. This interval specifies the two points $P(1, 9)$ and $Q(2, 2)$ indicated on the diagram.

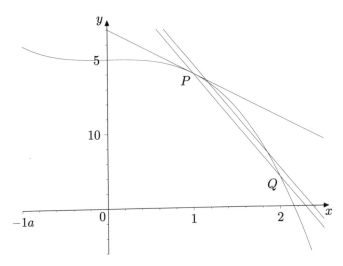

Figure 6.1   Linear approximations

Firstly we have drawn the tangent line at P. This approximates to the curve very well when x is near to 1, but as x increases towards 2 the line diverges from the curve by a considerable amount.

Secondly we have drawn the line passing through P and Q. This provides quite a good approximation across the interval, but near to P it is not so accurate as the tangent line.

Finally we have drawn a line parallel to the line PQ, for which the greatest difference between the y-value on the line and that on the curve, as x varies between 0 and 2, is smaller than it is for PQ. It is better in some places, but worse in others, especially near to $x = 1$ and $x = 2$.

Each of these lines is a possible candidate for a linear approximation. They have been chosen using different criteria, and this serves to emphasise that there is no "best" linear approximation.

Such considerations apply when we extend the discussion to approximating polynomials of higher degree than linear, and several different families of approximating polynomials have been investigated over the years.

In this chapter we shall base the choice of approximation on the notion of tangency, as with the first of the lines discussed above.

If we now generalise the situation so that the curve has equation $y = f(x)$, and the point P is $(a, f(a))$, then we can work out the equation of the tangent line, and it can be rearranged in the form

$$y = f(a) + f'(a)(x - a).$$

The tangent line and the curve have the same gradient at P, and so y changes as x changes at approximately the same rate along the line and the curve in a small interval containing $x = a$. This suggests that the tangent line will give a reasonable approximation to the curve near to $x = a$.

Even simpler than a linear approximation is to use a constant, so that we could say that near to $x = a$, $f(x)$ is approximately equal to $f(a)$. In the next section we shall consider the error which this constant approximation involves.

6.2 The Mean Value Theorem

A proof of this theorem is beyond the scope of this book. It can be found in Real Analysis books such as Howie, Chapter 4. It is important however to have a clear statement of the theorem with all the requisite conditions, and to give a geometrical interpretation.

Theorem 6.1 (The Mean Value Theorem)

Suppose that $f(x)$ is a function which is continuous for all values of x satisfying $a \leq x \leq b$, and differentiable for all values of x satisfying $a < x < b$. Then there exists a number c satisfying $a < c < b$ for which

$$\frac{f(b) - f(a)}{b - a} = f'(c).$$

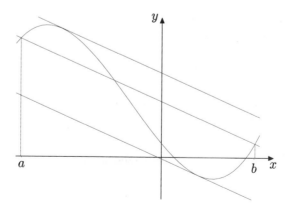

Figure 6.2   The Mean Value Theorem

Geometrically the left-hand side of the equation represents the slope of the chord joining the points $(a, f(a))$ and $(b, f(b))$. The right-hand side is the gradient of the tangent at the point $(c, f(c))$. So the conclusion of the theorem says that under the appropriate conditions of continuity and differentiability, given a chord joining two points on a graph, there is always a tangent at some intermediate point which is parallel to the chord. Figure 6.2 illustrates this, where we have shown a case when there is more than one possible value of c.

Example 6.2

Use the Mean Value Theorem to show that if $f'(x) > 0$ for all x satisfying $a < x < b$, then f is strictly increasing. What can we conclude if the condition $f'(x) > 0$ is replaced by $f'(x) \geq 0$?

Let x_1 and x_2 be any numbers satisfying $a \leq x_1 < x_2 \leq b$. Using the Mean Value Theorem gives
$$\frac{f(x_2) - f(x_1)}{x_2 - x_1} = f'(c) > 0.$$
Since the denominator $x_2 - x_1 > 0$ we deduce that $f(x_2) - f(x_1) > 0$. So f is an increasing function.

If the condition $f'(x) > 0$ is replaced by $f'(x) \geq 0$ the same reasoning tells us that $f(x_2) - f(x_1) \geq 0$. So f is still an increasing function, although not necessarily strictly increasing.

We can re-write the equation for the Mean Value Theorem in the form
$$f(b) = f(a) + f'(c)(b - a).$$

To interpret this as providing an approximation it is convenient to regard a as a fixed number, and b as a variable, which in accordance with usual notation we shall replace by x. We can then write
$$f(x) = f(a) + f'(c)(x - a),$$
where c lies somewhere between x and a.

This equation tells us that if the function $f(x)$ is approximated using the constant $f(a)$, then the error has the form $f'(c)(x - a)$. Geometrically this makes sense, because we should expect the error to depend on how far x is from a and on how large the rate of change of f is. Now in most cases we cannot find the value of c exactly. Indeed if we could, this would tell us the exact value of $f(x)$ and there would be no need to consider approximations at all.

Sometimes we can find estimates for f', usually in a form which tells us that for some real number M, and for all x satisfying $a \leq x \leq b$,
$$|f'(x)| \leq M.$$

We can then deduce that the approximation error satisfies

$$|f'(c)(x-a)| \le M|x-a|.$$

The right hand side is called an **error bound** for the constant approximation.

In order to study the error involved in using a linear approximation we need a version of the Mean Value Theorem applied to two functions simultaneously. We first note that if we apply to Mean Value Theorem to two functions f and g we obtain

$$\frac{f(b)-f(a)}{b-a} = f'(c_1); \quad \frac{g(b)-g(a)}{b-a} = g'(c_2).$$

If we divide these two equations the denominator $b-a$ cancels and we obtain

$$\frac{f(b)-f(a)}{g(b)-g(a)} = \frac{f'(c_1)}{g'(c_2)}.$$

There is no reason why c_1 and c_2 should be the same. However the next theorem shows that we can find a common value of c for the latter equation.

Theorem 6.3 (Generalised Mean Value Theorem)

If the two functions $f(x)$ and $g(x)$ are both continuous for $a \le x \le b$ and differentiable for $a < x < b$, and if $g'(x)$ is non-zero for all x satisfying $a < x < b$ then there is a number c between a and b such that

$$\frac{f(b)-f(a)}{g(b)-g(a)} = \frac{f'(c)}{g'(c)}.$$

Proof

We first comment that the denominator $g(b)-g(a)$ cannot be zero, for otherwise the Mean Value Theorem would tell us that $g'(c)=0$ for some c between a and b, contrary to the conditions of the theorem.

We now define the function $k(x)$ by means of the equation

$$k(x) = (f(b)-f(a))\,(g(x)-g(a)) - (g(b)-g(a))\,(f(x)-f(a)).$$

It is easy to see that $k(b) = k(a) = 0$, and that the conditions of the theorem ensure that $k(x)$ satisfies the conditions needed to apply the Mean Value Theorem. Therefore

$$0 = \frac{k(b)-k(a)}{b-a} = k'(c),$$

for some c between a and b. Differentiating $k(x)$ gives

$$k'(x) = (f(b) - f(a)) g'(x) - (g(b) - g(a)) f'(x),$$

and so we conclude that for some number c between a and b,

$$(f(b) - f(a)) g'(c) - (g(b) - g(a)) f'(c) = 0,$$

which can be rearranged in the form given in the statement of the theorem.   □

Now recall that the linear approximation we obtained in Section 6.1 had the equation $y = f(a) + f'(a)(x - a)$. So the error will be given by

$$E(x) = f(x) - (f(a) + f'(a)(x - a)).$$

We now apply the Generalised Mean Value Theorem with $f(x)$ replaced by $E(x)$ and with $g(x)$ replaced by $(x - a)^2$. Noting that $E(a) = 0$ and $g(a) = 0$ we deduce that for some number c between a and b

$$\frac{E(b)}{(b - a)^2} = \frac{E'(c)}{2(c - a)} = \frac{f'(c) - f'(a)}{2(c - a)} = \frac{f''(d)}{2},$$

where d is some number between a and c, obtained by applying the Mean Value Theorem to $f'(x)$ over the interval $a \leq x \leq c$.

We can summarise these results by saying that for some number d between a and x we have

$$f(x) = f(a) + f'(a)(x - a) + \frac{f''(d)}{2}(x - a)^2.$$

So the error term $\dfrac{f''(d)}{2}(x - a)^2$ depends on the second derivative, which tells us how fast the gradient of the graph of $f(x)$ is changing. If this is very large then the graph will quickly diverge from that of its linear approximation.

Sometimes we can find estimates for f'' which tell us that for some real number L, and for all x satisfying $a \leq x \leq b$,

$$|f''(x)| \leq L.$$

We can then deduce that the approximation error satisfies

$$\left| \frac{f''(d)}{2}(x - a) \right| \leq \frac{L}{2}|x - a|^2.$$

Now the error bound for the constant approximation involved $|x - a|$, whereas that for the linear involves $|x - a|^2$. So if for example $|x - a| \leq 10^{-3}$ then $|x - a|^2 \leq 10^{-6}$, and so it is possible that the linear approximation could have an accuracy of twice as many decimal places as the constant approximation (depending on the relative sizes of f' and f'').

Example 6.4

Find the equation of the tangent line to the curve whose equation is $y = \sqrt[3]{x}$ at the point $(8, 2)$, and use it to find an approximate value for $\sqrt[3]{8.2}$. Find an error bound and hence discuss the accuracy of the approximation.

To find the tangent and the error term we need the first two derivatives of $f(x) = \sqrt[3]{x}$. So we have

$$
\begin{aligned}
f(x) &= \sqrt[3]{x} = x^{\frac{1}{3}}; \\
f'(x) &= \frac{1}{3}x^{-\frac{2}{3}} = \frac{1}{3\left(\sqrt[3]{x}\right)^2} \\
f''(x) &= \frac{1}{3}\left(-\frac{2}{3}\right)x^{-\frac{5}{3}} = -\frac{2}{9\left(\sqrt[3]{x}\right)^5}.
\end{aligned}
$$

In this example we have $a = 8$, so the equation of the tangent line at $(8, 2)$ is

$$
y = 2 + \frac{1}{12}(x - 8).
$$

An approximation for $\sqrt[3]{8.2}$ is therefore

$$
2 + \frac{1}{12}(0.2) = 2 + \frac{1}{60} = 2.01666\ldots
$$

The error term is $\dfrac{f''(d)}{2}(0.2)^2$, where $8 < d < 8.2$.

Now f'' is a decreasing function of x for $x > 0$, so we have

$$
|f''(d)| \le |f''(8)| = \frac{2}{9\left(\sqrt[3]{8}\right)^5} = \frac{2}{9 \times 32}.
$$

So an error bound will be

$$
\frac{2}{9 \times 32} \times \frac{1}{2} \times (0.2)^2 = \frac{1}{9 \times 32 \times 25} = \frac{1}{7200} = 0.00013888\ldots
$$

Now f'' is negative for $x > 0$, so the error is negative. Therefore

$$
2.01666\ldots - 0.00013888\ldots \le \sqrt[3]{8.2} \le 2.01666\ldots,
$$

which tells us that $2.01652 < \sqrt[3]{8.2} < 2.01667$. Therefore we can say that $\sqrt[3]{8.2} \approx 2.017$ with an error of less than 1 in the third decimal place.

Example 6.5

Use the Generalised Mean Value Theorem to give a proof of l'Hôpital's Rule. (See Section 2.5.4.)

Suppose that $f(x)$ and $g(x)$ satisfy the conditions of l'Hôpital's rule, and that $\dfrac{f'(x)}{g'(x)} \to l$ as $x \to a$. We apply the Generalised Mean Value Theorem, noting that $f(a) = g(a) = 0$. Therefore

$$\frac{f(x)}{g(x)} = \frac{f(x) - f(a)}{g(x) - g(a)} = \frac{f'(c)}{g'(c)} \to l \text{ as } x \to a,$$

because $a < c < x$ and so $c \to a$ as $x \to a$. This completes the proof of l'Hôpital's Rule.

6.3 Quadratic Approximation

We saw in section 6.1 that a linear approximation using the tangent line is quite good in a small neighbourhood of the point of contact, but less good the further away we are from this point. To improve the approximation we replace the straight line by a curve whose gradient changes in such a way as to try to follow the graph of $f(x)$, the function being approximated.

After a linear approximation the next simplest would be a quadratic. In the case of the tangent line, it passed through the point of contact and had the same gradient as the curve at that point. We shall choose the quadratic passing through the point of contact $(a, f(a))$ and having the same first and second derivatives as $f(x)$ at $x = a$.

Suppose that the quadratic has equation $y = px^2 + qx + r$. Using the above conditions will give three equations, which will be sufficient to determine the coefficients of the quadratic. We therefore require

$$
\begin{aligned}
f''(a) &= 2p, \text{ giving } p = \frac{f''(a)}{2}; \\
f'(a) &= 2pa + q, \text{ giving } q = f'(a) - af''(a); \\
f(a) &= pa^2 + qa + r.
\end{aligned}
$$

If we now substitute for p and q from the first two equations into the third, and then rearrange the result we can obtain

$$r = f(a) - af'(a) + \frac{a^2}{2} f''(a).$$

We have therefore found the coefficients for the approximating quadratic, whose equation can now be rearranged in the form

$$y = f(a) + f'(a)(x - a) + \frac{f''(a)}{2}(x - a)^2.$$

We see that the first two terms are those which comprise the linear approximation, and this suggests that we have now obtained the first two in a sequence of such approximating polynomials of increasing degree. We shall pursue this in the next section.

A similar error analysis is possible to that for constant and linear approximations. We shall however consider this in the more generalised context of polynomial approximations of degree n in the next section.

6.4 Taylor Polynomials

This family of polynomial approximations was investigated by the English mathematician Brook Taylor (1685–1731) and by the Scottish mathematicians Colin Maclaurin (1698–1756) and James Gregory (1638–1675).

We shall begin by investigating an expansion for a function in the form of an infinite series of powers which we shall then use to generate approximations near to $x = a$. Generalising the form of linear and quadratic approximations we assume a series of the form

$$f(x) = a_0 + a_1(x - a) + a_2(x - a)^2 + a_3(x - a)^3 + \cdots$$

If we substitute $x = a$ then all the terms are zero except the first, giving $f(a) = a_0$, which tells us what the constant term must be in terms of f. This is the aim for all the coefficients in the series, to find them in terms of f. We shall assume that it is valid to differentiate this infinite series term-by-term, to obtain

$$f'(x) = a_1 + 2a_2(x - a) + 3a_3(x - a)^2 + 4a_4(x - a)^3 + \cdots$$

Substituting $x = a$ then gives $f'(a) = a_1$. Differentiating again gives

$$f''(x) = 2a_2 + 3.2a_3(x - a) + 4.3a_4(x - a)^2 + \cdots,$$

and now substituting $x = a$ gives $f''(a) = 2a_2$.

If we repeat this process we find that $f'''(a) = 3.2a_3$, $f^{(4)}(a) = 4.3.2a_4$, and so on. This generalises to suggest that $f^{(n)}(a) = n!a_n$. So we have formulae for the coefficients in terms of f and its derivatives.

Definition 6.6

We define the **Taylor series expansion of $f(x)$ about $x = a$** to be the series

$$f(a) + f'(a)(x - a) + \frac{f''(a)}{2!}(x - a)^2 + \cdots + \frac{f^{(n)}(a)}{n!}(x - a)^n + \cdots$$

When we truncate this series we obtain the **Taylor polynomial of** $f(x)$ **about** $x = a$ **of degree** n, denoted by

$$P_{n,a}(x) = f(a) + f'(a)(x-a) + \frac{f''(a)}{2!}(x-a)^2 + \cdots + \frac{f^{(n)}(a)}{n!}(x-a)^n.$$

The special case $a = 0$ is called the **Maclaurin expansion**, which is therefore a series involving just powers of x,

$$f(0) + f'(0)x + \frac{f''(0)}{2!}x^2 + \cdots + \frac{f^{(n)}(0)}{n!}x^n + \cdots$$

In finding Taylor series we shall assume some results from Real Analysis:

1. The Taylor series usually converges to the function $f(x)$ that we start with, but there are exceptions.

2. We can perform algebraic operations on series to get new series, such as addition, multiplication, substitution, differentiation and integration.

3. If a function is equal to some series expansion in powers of x (or $x - a$) then that must be the Maclaurin (Taylor) series, i.e., the series is unique.

4. There are precise conditions for the validity of these results, discussed in Howie, Chapters 2, 4 and 7. The examples we shall consider have been chosen to satisfy those conditions.

5. Certain basic expansions for elementary functions form part of school mathematics and should be memorised.

$$e^x = 1 + x + \frac{x^2}{2!} + \frac{x^3}{3!} + \cdots + \frac{x^n}{n!} + \cdots$$

$$\sin x = x - \frac{x^3}{3!} + \frac{x^5}{5!} + \cdots + (-1)^n \frac{x^{2n+1}}{(2n+1)!} + \cdots$$

$$\cos x = 1 - \frac{x^2}{2!} + \frac{x^4}{4!} + \cdots + (-1)^n \frac{x^{2n}}{(2n)!} + \cdots$$

Example 6.7

Find the Maclaurin expansion of $\dfrac{1}{1-x}$.

We can recognise $\dfrac{1}{1-x}$ as the sum of an infinite Geometric Series, or we can expand $(1-x)^{-1}$ by the Binomial Theorem. In either case we obtain

$$\frac{1}{1-x} = 1 + x + x^2 + x^3 + \cdots + x^n + \cdots,$$

valid for $|x| < 1$. Property 3 in the list above tells us that this must be the Maclaurin series.

Example 6.8

Find the Maclaurin expansion of $\dfrac{1}{1+x}$.

We can use the result of Example 6.7 together with property 2 above, replacing x with $-x$ to obtain

$$\frac{1}{1+x} = 1 - x + x^2 - x^3 + \cdots + (-1)^n x^n + \cdots,$$

valid for $|x| < 1$, and Property 3 tells us that this is the Maclaurin series. This is also a useful expansion to memorise.

Example 6.9

Find the Maclaurin series for $\ln(1+x)$.

We use the result of Example 6.8 and integrate both sides. We use the fact that $\ln 1 = 0$ to show that the constant of integration is zero. Therefore

$$\ln(1+x) = x - \frac{x^2}{2} + \frac{x^3}{3} + \cdots + (-1)^{n+1}\frac{x^n}{n} + \cdots.$$

Example 6.10

Find the Maclaurin expansion for $\tan^{-1} x$.

We begin with

$$\frac{1}{1+x} = 1 - x + x^2 - x^3 + \cdots + (-1)^n x^n + \cdots,$$

and then substitute $x = t^2$ to obtain

$$\frac{1}{1+t^2} = 1 - t^2 + t^4 - t^6 + \cdots + (-1)^n t^{2n} + \cdots.$$

We now integrate both sides to obtain

$$\int_0^x \frac{1}{1+t^2}\, dt = \left[t - \frac{t^3}{3} + \frac{t^5}{5} + \cdots + (-1)^n \frac{t^{2n+1}}{2n+1} + \cdots \right]_0^x.$$

Therefore

$$\tan^{-1} x = x - \frac{x^3}{3} + \frac{x^5}{5} + \cdots + (-1)^n \frac{x^{2n+1}}{2n+1} + \cdots.$$

Example 6.11

Find the Maclaurin expansion for $\cosh x$.

Using the definition of cosh and the known expansion for the exponential function gives

$$
\begin{aligned}
\cosh x &= \frac{e^x + e^{-x}}{2} \\
&= \frac{1}{2}\left(1 + x + \frac{x^2}{2!} + \frac{x^3}{3!} + \cdots + \frac{x^n}{n!} + \cdots\right) \\
&\quad + \frac{1}{2}\left(1 - x + \frac{x^2}{2!} - \frac{x^3}{3!} + \cdots + (-1)^n \frac{x^n}{n!} + \cdots\right) \\
&= 1 + \frac{x^2}{2!} + \frac{x^4}{4!} + \cdots + \frac{x^{2n}}{(2n)!} + \cdots.
\end{aligned}
$$

We have seen that in none of the examples above have we used the formula for the coefficients in terms of the derivatives of the function. This is often much more complicated than using a known expansion and applying various operations. In the next example we compare the two methods.

Example 6.12

Find the Maclaurin expansion for $f(x) = \exp\left(-x^2\right)$.

If we try to find successive derivatives, with the aim of finding a formula for the n-th derivative, we find that

$$
\begin{aligned}
f(x) &= \exp\left(-x^2\right) \\
f'(x) &= -2x \exp\left(-x^2\right) \\
f''(x) &= -2 \exp\left(-x^2\right) + 4x^2 \exp\left(-x^2\right) \\
f'''(x) &= 12x \exp\left(-x^2\right) - 8x^3 \exp\left(-x^2\right)
\end{aligned}
$$

We can see that this will become complicated, and as with previous examples it is more straightforward to use the known expansion for the exponential function and then use substitution, as follows.

$$
\begin{aligned}
\exp t &= 1 + t + \frac{t^2}{2!} + \frac{t^3}{3!} + \frac{t^4}{4!} + \cdots \\
\exp(-t) &= 1 - t + \frac{t^2}{2!} - \frac{t^3}{3!} + \frac{t^4}{4!} - \cdots \\
\exp\left(-x^2\right) &= 1 - x^2 + \frac{x^4}{2!} - \frac{x^6}{3!} + \frac{x^8}{4!} - \cdots
\end{aligned}
$$

Example 6.13

Find the Taylor expansion for $\cos x$ about $x = \dfrac{\pi}{2}$.

We shall show two methods. In the first we calculate the successive derivatives, and in the second we use a known expansion.

Method 1

The following table shows the successive derivatives, and in the right-hand
column their values when $x = \dfrac{\pi}{2}$.

$f(x)$	$\cos x$	0
$f'(x)$	$-\sin x$	-1
$f''(x)$	$-\cos x$	0
$f'''(x)$	$\sin x$	1
$f^{(4)}(x)$	$\cos x$	0
$f^{(5)}(x)$	$-\sin x$	-1
$f^{(6)}(x)$	$-\cos x$	0

So we can write down the Taylor series as follows.

$$f\left(\frac{\pi}{2}\right) \;+\; f'\left(\frac{\pi}{2}\right)\left(x - \frac{\pi}{2}\right) + \frac{f''\left(\frac{\pi}{2}\right)}{2!}\left(x - \frac{\pi}{2}\right)^2 + \frac{f'''\left(\frac{\pi}{2}\right)}{3!}\left(x - \frac{\pi}{2}\right)^3$$

$$+\; \frac{f^{(4)}\left(\frac{\pi}{2}\right)}{4!}\left(x - \frac{\pi}{2}\right)^4 + \frac{f^{(5)}\left(\frac{\pi}{2}\right)}{5!}\left(x - \frac{\pi}{2}\right)^5 + \cdots$$

$$=\; -\left(x - \frac{\pi}{2}\right) + \frac{1}{3!}\left(x - \frac{\pi}{2}\right)^3 - \frac{1}{5!}\left(x - \frac{\pi}{2}\right)^5 + \cdots.$$

Method 2

The known expansions we have are Maclaurin expansions, about $x = 0$, so
we need to perform a translation by putting $y = x - \dfrac{\pi}{2}$. In this way when $x = \dfrac{\pi}{2}$
we shall have $y = 0$.

Using this transformation $\cos x$ becomes $\cos\left(y + \dfrac{\pi}{2}\right) = -\sin y$, using a
trigonometric identity. The expansion of $-\sin y$ about $y = 0$ is

$$-y + \frac{y^3}{3!} - \frac{y^5}{5!} + \cdots.$$

Therefore the expansion for $\cos x$ about $x = \dfrac{\pi}{2}$ is

$$-\left(x - \frac{\pi}{2}\right) + \frac{1}{3!}\left(x - \frac{\pi}{2}\right)^3 - \frac{1}{5!}\left(x - \frac{\pi}{2}\right)^5 + \cdots.$$

Example 6.14

We can use the Maclaurin series for the exponential function to prove the result that the exponential function tends to infinity faster than any power. This was discussed in Example 2.29, but is sufficiently important to be considered here also.

Using the Maclaurin expansion for positive values of x we have

$$\mathrm{e}^x = 1 + x + \frac{x^2}{2!} + \frac{x^3}{3!} + \cdots + \frac{x^n}{n!} + \frac{x^{n+1}}{(n+1)!} + \cdots > \frac{x^{n+1}}{(n+1)!}.$$

We therefore have, for any n,

$$0 < \frac{x^n}{\mathrm{e}^x} < x^n \frac{(n+1)!}{x^{n+1}} = \frac{(n+1)!}{x} \to 0 \text{ as } x \to \infty.$$

Therefore by squeezing (Section 2.5.1) we have shown that

$$\lim_{x \to \infty} \frac{x^n}{\mathrm{e}^x} = 0.$$

6.5 Taylor's Theorem

Theorem 6.15 (Taylor's Theorem)

If the $(n+1)$-th derivative of f exists throughout an interval containing a and x, and if $P_{n,a}(x)$ denotes the Taylor polynomial of $f(x)$ of degree n about the point a, then we have

$$f(x) = P_{n,a}(x) + E_n(x),$$

where the error (or remainder) term $E_n(x)$ is given by

$$E_n(x) = \frac{f^{(n+1)}(c)}{(n+1)!}(x-a)^{n+1},$$

where c is some number between a and x. This is known as Lagrange's form of the remainder, named after the French mathematician Joseph-Louis Lagrange (1736–1813).

We shall not give a proof of Taylor's Theorem in this book. Proofs can be found in Real Analysis textbooks, for example in Howie Chapter 4.

As we remarked in Section 6.4, we can use the Taylor expansion to obtain polynomial approximations for a function by truncating the series after a finite number of terms. As with the constant and linear approximations, we need to be able to find error bounds, and the error term in Taylor's Theorem enables us to do this.

Example 6.16

Use Taylor's Theorem to calculate an approximate value for $\ln 1.05$, accurate to six places of decimals.

We consider the expansion of $\ln(1+x)$ about $x = 0$, which we found in Example 6.9. However to investigate the error we need to find an expression for the $(n+1)$-th derivative. In the table below we have calculated the first few derivatives, to the extent that we can see a pattern and write down an expression for the $(n+1)$-th derivative. Strictly speaking we should give a proof of our formula, by mathematical induction, but we shall omit that here. Beginning with $f(x)$ and repeatedly differentiating gives

$$
\begin{aligned}
f(x) &= \ln(1+x) \\
f'(x) &= (1+x)^{-1} \\
f''(x) &= -(1+x)^{-2} \\
f'''(x) &= 2(1+x)^{-3} \\
f^{(4)}(x) &= -3 \times 2(1+x)^{-4} \\
f^{(5)}(x) &= 4 \times 3 \times 2(1+x)^{-5} \\
&\vdots \\
f^{(n)}(x) &= (-1)^{n-1}(n-1)!(1+x)^{-n} \\
f^{(n+1)}(x) &= (-1)^{n}n!(1+x)^{-(n+1)}.
\end{aligned}
$$

The error term is then given by

$$
E_n(x) = \frac{(-1)^n n!}{(n+1)!} x^{n+1}(1+c)^{-(n+1)} = \frac{(-1)^n}{(n+1)} x^{n+1}(1+c)^{-(n+1)}.
$$

To find the approximation for $\ln(1.05)$ we put $x = 0.05$. Therefore $0 < c < 0.05$ and so we know that $(1+c) > 1$, telling us that $\dfrac{1}{1+c} < 1$, and hence

$$
|E_n(0.05)| \le \frac{(0.05)^{n+1}}{n+1}.
$$

To achieve six decimal places of accuracy we need this error bound to be sufficiently small, and using a calculator gives error bounds for $n = 1, 2, 3, 4$ as follows.

n	Error bound
1	1.25×10^{-3}
2	4.167×10^{-5}
3	1.5625×10^{-6}
4	6.25×10^{-8}

So $n = 3$ will not give sufficient accuracy, but $n = 4$ will. This means that we should use the polynomial approximation

$$\ln(1+x) \approx x - \frac{x^2}{2} + \frac{x^3}{3} - \frac{x^4}{4},$$

$$\ln(1.05) \approx 0.05 - \frac{0.05^2}{2} + \frac{0.05^3}{3} - \frac{0.05^4}{4}.$$

Working this out on a calculator gives a readout of 0.048790104 and the calculator manual suggests that no more than the last digit would be suspect.

Now with $n = 4$ the error term is positive, and so we can say that

$$0.04879010 < \ln(1.05) < 0.04879011 + 6.25 \times 10^{-8} < 0.04879018.$$

So we can be certain that $\ln(1.05) \approx 0.048790$ to six decimal places of accuracy.

We can see in this example that not only do we have to take the error bound into account, but also any possible errors arising from the use of a calculator or a computer.

6.6 Using MAPLE for Taylor Series

We can use MAPLE both algebraically and geometrically to investigate Taylor series. Firstly there is a built-in command which works out Taylor series, in which we have to specify the function, the point about which the expansion is to take place, and how many terms we want.

Example 6.17

Find the first five terms of the Taylor series used in Example 6.16 using MAPLE.

We use the command `taylor(ln(1+x),0,5);`
This produces as its output

$$x - \frac{x^2}{2} + \frac{x^3}{3} - \frac{x^4}{4} + O(x^5).$$

The last part of the expression represents the fact that the order of magnitude, symbolised by the capital O, of the error term is x^5, in accordance with Taylor's Theorem in which $E_n(x)$ contains $(x - a)^{n+1}$.

Notice that although we asked for five terms there appear only to be four. This is because the constant term in this expansion is zero.

To investigate the successive approximations we can plot them on the same graph, using the command

```
plot([ln(1+x),x,x-x^/2,x-x^2/2+x^3/3,x-x^2/2+x^3/3-x^4/4],
x=0..2,color=black);
```

A command such as this would become very long if we wanted to plot 10 approximations for example. Where we have a regular pattern which tells us one term in terms of the previous ones MAPLE contains a built-in programming language which would enable us to do this.

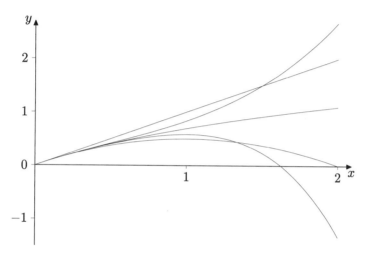

Figure 6.3   Taylor approximations for $\ln(1 + x)$

Figure 6.3 contains the graph of $\ln(1 + x)$ and the linear, quadratic, cubic and quartic approximations. MAPLE does not label its graphs on a plot such as this, so we have to inspect each individual graph to decide which of them corresponds to each approximation, which is a good exercise in itself. In fact MAPLE uses different colours for each graph in a set such as this (we have included the command `color=black` solely for printing purposes). This makes it easier to identify each graph. In fact the `color` command can be adapted to specify the colours we wish for each of the approximations.

In this example the series is a Maclaurin series, but MAPLE can expand about any point.

Example 6.18

Use MAPLE to find the Taylor expansion for $\cos x$ about $x = \dfrac{\pi}{2}$.

This is the series we found in Example 6.13. To produce the same number of terms as we displayed in that example we need the command

$$\texttt{taylor(cos(x),x=Pi/2,6);}$$

This produces the output

$$-\left(x - \frac{\pi}{2}\right) + \frac{1}{6}\left(x - \frac{\pi}{2}\right)^3 - \frac{1}{120}\left(x - \frac{\pi}{2}\right)^5 + O\left(\left(x - \frac{\pi}{2}\right)^6\right),$$

where the only difference from the result in Example 6.13 is that MAPLE works out the arithmetic values of the factorials.

EXERCISES

6.1. Find the equation of the tangent line to the curve whose equation is $y = \sqrt{x}$ at the point $(9, 3)$, and use it to find an approximate value for $\sqrt{9.01}$. Find an error bound and hence discuss the accuracy of the approximation.

6.2. Find the equation of the tangent line to the curve whose equation is $y = \tan x$ at the point $\left(\frac{\pi}{4}, 1\right)$, and use it to find an approximate value for $\tan 47°$. Find an error bound and hence discuss the accuracy of the approximation.

6.3. Use the Mean Value Theorem to show that if a function has a negative derivative throughout an interval, then the function is decreasing throughout that interval.

6.4. Find the quadratic approximation for e^x at $x = 2$. Use a calculator or a spreadsheet to compare e^x with the quadratic approximation for values of x at intervals of 0.1 between 0 and 4. Plot e^x and the quadratic approximation on the same graph, for $0 \le x \le 4$ using MAPLE or a graphical calculator.

6.5. Find the Maclaurin expansions for the following functions.

(a) $\sin\left(3x^2\right)$; (b) $\ln\left(2 - x^2\right)$;

(c) $\exp\left(1 + x^3\right)$; (d) $\sqrt{x + 2}$;

(e) $\dfrac{1}{\left(1 + x^2\right)^2}$; (f) $\dfrac{1 + x^2}{1 - x}$;

(g) $\cos 2x$; (h) $\sin^2 x$;

(i) $\dfrac{\sin x}{x}$; (j) $\cos\left(x - \dfrac{\pi}{4}\right)$;

(k) $\sinh(x^3)$; (l) $\cosh x - \cos x$.

6.6. Find the Taylor series for the following functions about the points indicated.

(a) e^x about $x = 1$; (b) $\sin x$ about $x = -\pi$;

(c) $\sqrt[3]{x}$ about $x = 2$; (d) $\ln x$ about $x = 3$;

(e) $\dfrac{1}{x} + \dfrac{1}{x^2}$ about $x = -1$; (f) $\dfrac{x}{1 - x}$ about $x = -1$;

(g) $\cosh x$ about $x = -2$; (h) $x \sin x$ about $x = \pi$;

(i) $\ln(2 + x)$ about $x = 2$; (j) e^{x+3} about $x = 2$.

6.7. Find the Maclaurin expansion for $\ln(2 + 3x)$ and find an expression for the error term $E_n(x)$.

6.8. Find the Maclaurin expansion for $\sqrt{4 + x}$ as far as the term involving x^4. Find an expression for the error term $E_4(x)$, and find an error bound when $x = 0.1$.

6.9. Use Taylor's Theorem to find an approximation for $\cos(5^\circ)$, correct to six decimal places. (Use $E_n(x)$ for even values of n.)

7
Integration

Ideas about integration have been around much longer than those of differentiation. The Greek mathematician Archimedes (3rd Century B.C.) knew how to calculate the area of a segment of a parabola by "quadrature", which involved approximation by regions of known area such as quadrilaterals or triangles. Integration as the reverse of differentiation was a much later idea, after the invention of the differential calculus in the time of Newton (17th Century A.D.). Bringing the two views of integration together was the work of mathematicians in the 19th Century in particular, culminating in the work of Bernhard Riemann (1826–1866), whose name is associated with the theory of integration which is often studied in Real Analysis courses (see Howie Chapter 5 for example).

In this chapter we shall review some of the basic ideas about integration, briefly revise some standard results from school calculus, and develop some aspects of integration using limits.

7.1 Integration as Summation

The basic idea is that of "area under a graph", where this phrase is commonly used to signify the area of a plane region bounded by the graph of a non-negative function $y = f(x)$, the x-axis, and the lines $x = a$ and $x = b$. We approximate this area using rectangles, from below and from above. These ideas are illustrated in Figure 7.1. For functions that we commonly use, with continuous graphs for example, it turns out that the approximations from below and from above can be made as close as we wish. As these approximations get better and better they "home in" on a definite numerical value which corresponds to the area of the region. This is a limiting process, but a more complicated one than we discussed in Chapter 2.

We generate the approximations by subdividing the interval $a \leq x \leq b$ by means of an increasing sequence of points

$$a = x_1 < x_2 < \cdots < x_{n-1} < x_n = b.$$

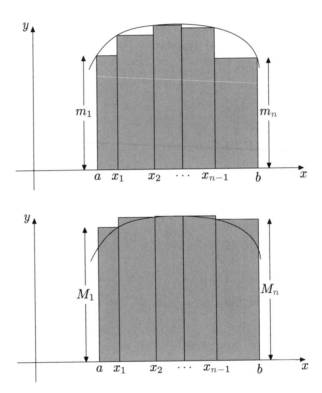

Figure 7.1   The integral as a sum

The sum of the areas of the rectangles lying below the graph (as in the top diagram) is called the **lower sum** corresponding to the subdivision, and the sum of the areas of the rectangles enclosing the area underneath the graph (as in the bottom diagram) is called the **upper sum** corresponding to the subdivision. We introduce the notation for these as follows.

$$s = \sum_{i=1}^{n} m_i \left(x_i - x_{i-1} \right); \quad S = \sum_{i=1}^{n} M_i \left(x_i - x_{i-1} \right).$$

In these sums each term represents the area of a rectangle, where $(x_i - x_{i-1})$ represents the width and m_i or M_i represents the height, in the first and second sum respectively.

The fundamental idea is that as the lengths of the intervals of the subdivision all tend to zero, both the upper and lower sum tend to a common limit, the area under the curve.

The theoretical underpinning of this idea is quite extensive, and beyond the scope of this book. It can be shown that for "respectable" functions (given by standard formulae for example) we can use any convenient approximating sum in this way, where we replace M_i or m_i with any number between, usually of the form $f(c_i)$, where $x_{i-1} \le c_i \le x_i$. This gives the sum

$$\sum_{i=1}^{n} f(c_i)\,(x_i - x_{i-1}).$$

This sum also represents a sum of areas of rectangles. The theory shows that all such sums have a common limit as the number of intervals in the subdivision increases, and their lengths all tend to zero. This common limit is called the integral, denoted by

$$\int_{a}^{b} f(x)\,dx,$$

and a function for which this common limit exists is said to be **integrable** over the interval $a \le x \le b$. The function $f(x)$ which we are integrating is referred to as the **integrand**. We shall utilise the idea of the integral as the limit of a sum in Chapter 11, applied to geometrical quantities such as area and volume.

The connection between this process and differentiation was established by means of the Fundamental Theorem of Calculus, which says that under appropriate conditions

$$\int_{a}^{b} f'(x)\,dx = f(b) - f(a).$$

The first clear statement of this theorem is attributed to Isaac Newton in some unpublished work written in 1666.

7.2 Some Basic Integrals

In the remainder of this chapter, and in Chapters 8, 9 and 10, we shall consider integration as the reverse of differentiation. Given a function $f(x)$, the problem is to find another function $F(x)$ whose derivative is $f(x)$. Such a function $F(x)$

is called an **indefinite integral**, denoted by $\int f(x)\,dx$. In contrast, the integral

$\int_a^b f(x)\,dx$ discussed above is called a **definite integral**.

We first note that if $\dfrac{d}{dx}F(x) = f(x)$ then $\dfrac{d}{dx}(F(x) + C) = f(x)$ for any real number C.

C is known as the constant of integration, and strictly speaking should be included whenever we evaluate an indefinite integral.

Basic integration is normally first encountered in school mathematics, and it is assumed that readers are familiar with a small number of indefinite integrals, as in the following table, where we have omitted the constant of integration, as we shall do throughout this book.

$f(x)$	$\int f(x)\,dx$
$x^\alpha \ \ (\alpha \neq -1)$	$\dfrac{x^{\alpha+1}}{\alpha + 1}$
x^{-1}	$\ln|x|$
$\cos x$	$\sin x$
$\sin x$	$-\cos x$
$\sec^2 x$	$\tan x$
e^{kx}	$\dfrac{e^{kx}}{k}$

Various rules and identities can be used to reduce many integrals to the basic ones above. There are some standard methods for doing this which are applicable to various classes of functions, and we shall consider these in later chapters. The examples in this section illustrate this idea with some elementary integrals. The algebraic rules of integration used in this chapter are as follows.

$$\int (Cf_1(x) + Df_2(x))\,dx = C\int f_1(x)\,dx + D\int f_2(x)\,dx,$$

where C and D are constants (the addition rule), and

$$\int_a^b f(x)\,dx = \int_a^c f(x)\,dx + \int_c^b f(x)\,dx.$$

Example 7.1

Evaluate $\int \cos 2x\,dx$.

The table above suggests that the answer will involve $\sin 2x$, and we can check by differentiation.

Now $\dfrac{d}{dx} \sin 2x = 2 \cos 2x$, and so we can see that we have to compensate for the factor of 2. We can therefore write $\dfrac{d}{dx} \dfrac{\sin 2x}{2} = \cos 2x$, and so

$$\int \cos 2x \, dx = \frac{\sin 2x}{2}.$$

A Common Mistake

In Example 7.1 we are integrating a function of the form $\cos\left(f(x)\right)$, where in this case $f(x) = 2x$. It looks as if we have written $\sin\left(f(x)\right)$ and then divided by the derivative of $f(x)$. This is NOT a general rule. It only works in this case because $f(x)$ is a linear function. We can see that the general rule is WRONG by the following example.

It is NOT TRUE that $\displaystyle\int \cos\left(x^2\right) \, dx = \dfrac{\sin\left(x^2\right)}{2x}$. We can check this by differentiation. We need to use the quotient rule, and we then find that

$$\frac{d}{dx}\left(\frac{\sin\left(x^2\right)}{2x}\right) = \frac{2x \times 2x \cos\left(x^2\right) - 2\sin\left(x^2\right)}{(2x)^2},$$

which is clearly NOT equal to $\cos\left(x^2\right)$. The moral of this example is therefore

Always check your answer by differentiation.

Example 7.2

Evaluate $\displaystyle\int \sin^2 x \, dx$.

Many trigonometric integrals involve the use of identities. In this case the double angle formula $\cos 2x = 1 - 2\sin^2 x$ is the one which is relevant. Therefore

$$\int \sin^2 x \, dx = \int \frac{1}{2}(1 - \cos 2x) \, dx = \frac{1}{2}\left(x - \frac{\sin 2x}{2}\right).$$

Example 7.3

Evaluate $\displaystyle\int \cos 5x \cos 3x \, dx$.

The last identity in Example 1.24 enables us to convert the product of cosines into a sum. We have $\cos 5x \cos 3x = \frac{1}{2}(\cos 8x + \cos 2x)$, so that

$$\int \cos 5x \cos 3x \, dx = \frac{1}{2}\int (\cos 8x + \cos 2x) \, dx = \frac{\sin 8x}{16} + \frac{\sin 2x}{4}.$$

Example 7.4

Evaluate $\int a^x \, dx \ (a > 0)$.

We recall Definition 1.27, telling us that $a^x = e^{x \ln a}$. Using this, and realising that $\ln a$ is simply a constant, we can write

$$\int a^x \, dx = \int e^{x \ln a} \, dx = \frac{e^{x \ln a}}{\ln a} = \frac{a^x}{\ln a}.$$

A Common Mistake

A common error in Example 7.4 is to treat the variable x as if it were a constant, writing $\int a^x \, dx = \dfrac{a^{x+1}}{x+1}$, which is WRONG. The mistake is demonstrated if we take the special case where $a = e$, in which case the expected answer is simply e^x and not $\dfrac{e^{x+1}}{x+1}$. What has been done in making this error is effectively to integrate with respect to a instead of with respect to x.

In the following example we shall interpret the integral of a negative function as the area between the graph and the x-axis, but with a negative sign. We therefore need to explain this briefly.

Suppose $f(x) \leq 0$ for all x in the interval $a \leq x \leq b$. Then if we let $g(x) = -f(x)$, we have $g(x) \geq 0$ for all x in the interval. So $\int_a^b g(x) \, dx$ is equal to the area A between the graph of the non-negative function $g(x)$ and the x-axis, bounded by the lines $x = a$ and $x = b$. Using the addition rule for integrals with $C = -1$ and $D = 0$ gives

$$A = \int_a^b g(x) \, dx = \int_a^b -f(x) \, dx = -\int_a^b f(x) \, dx \ \text{ so } \ \int_a^b f(x) \, dx = -A.$$

In terms of the upper and lower sums introduced in Section 7.1, since $f(x) \leq 0$ it follows that $m_i \leq 0$ and $M_i \leq 0$, so that each sum corresponds to the negative of the total area of the approximating rectangles. Again this will tell us that

$$\int_a^b f(x) \, dx = -A.$$

Example 7.5

Evaluate $\int_{-3}^3 \ln \left(1 + x^2\right) \sin x \, dx$.

The function looks complicated to integrate as an indefinite integral, so we have to look at it another way, in this case geometrically. We notice that the function is an odd function, and the interval of integration is symmetric about the origin. The answer is therefore zero. This is clear from Figure 7.2, with the interpretation explained above, that areas below the x-axis correspond to a negative answer for the integral. We have implicitly used the rule of integration telling us that

$$\int_{-3}^{3} \ln\left(1 + x^2\right) \sin x \, dx = \int_{-3}^{0} \ln\left(1 + x^2\right) \sin x \, dx + \int_{0}^{3} \ln\left(1 + x^2\right) \sin x \, dx.$$

The value of the first integral on the right-hand side is then minus that of the second integral, so they cancel to zero.

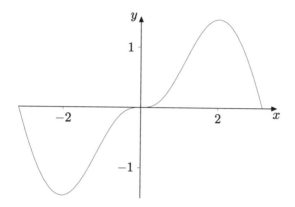

Figure 7.2   Integral of an odd function

Example 7.6

Evaluate $\displaystyle\int_{0}^{2} \sqrt{2x - x^2} \, dx$.

As in the previous example, we can interpret this integral as an area and evaluate it without calculation. The graph of the integrand is shown in Figure 7.3, and it appears to be a semicircle. We can verify this algebraically. The equation of the graph can be rearranged as follows.

$$y = \sqrt{2x - x^2} = \sqrt{1 - x^2 + 2x - 1} = \sqrt{1 - (x - 1)^2},$$

and therefore

$$(x - 1)^2 + y^2 = 1,$$

which is the equation of a circle centered at $(1, 0)$ with radius 1. The original equation involves the square root, and corresponds to the upper half of this circle.

The value of the integral is therefore equal to the area of the semicircular region, namely $\pi/2$.

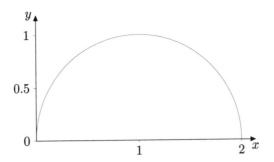

Figure 7.3   Graph for Example 7.6

7.3 The Logarithmic Integral

One of the items in the table of integrals at the beginning of Section 7.2 states that $\int \dfrac{1}{x}\,dx = \ln|x|$. Some textbooks say that $\int \dfrac{1}{x}\,dx = \ln x + C$. Neither of these is strictly correct, and in this section we shall discuss this integral further. The problem is caused by the fact that neither $1/x$ nor $\ln|x|$ is defined for $x = 0$. We approach the problem through graphs.

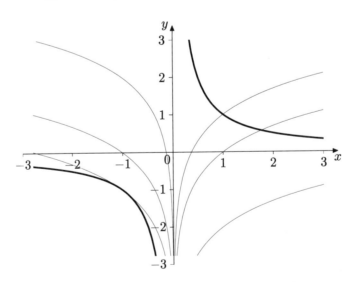

Figure 7.4   The logarithmic integral

In Figure 7.4 the thicker graph is that of $y = \dfrac{1}{x}$, and the other graphs are of $\ln x + C$ for $x > 0$, and $\ln(-x) + D$ for $x < 0$. All these logarithmic graphs have derivative $1/x$. For $x > 0$ this is a standard result. Now $1/x$ is an odd function, and so its gradient at $-x$ will be the negative of the gradient at x. This means that for $x < 0$, $1/x$ is the gradient for the logarithmic function with x replaced by $-x$, i.e., $\ln(-x)$. For $x < 0$ we can also verify this result using the chain rule.

$$\frac{d}{dx}(\ln(-x) + D) = \frac{1}{-x}\frac{d}{dx}(-x) = \frac{1}{-x}.(-1) = \frac{1}{x}.$$

Finally we have to consider the constant of integration. We can see that we could choose an arbitrary function of the form $\ln x + C$ for $x > 0$, and an arbitrary function of the form $\ln(-x) + D$ for $x < 0$. There is no reason why C should be the same as D, since the two "halves" do not join together, because of discontinuity at $x = 0$. So the most complete description of the set of functions whose derivative is $1/x$ is

$$F(x) = \begin{cases} \ln(x) + C & (x > 0), \\ \ln(-x) + D & (x < 0). \end{cases}$$

To write $\ln|x| + C$ is a convenient abbreviation, but it conceals the fact that C and D can be different.

7.4 Integrals with Variable Limits

So far when we have evaluated a definite integral of the form $\displaystyle\int_a^b f(x)\,dx$, the limits of integration a and b have been constants. There is no reason however why they should not involve a variable.

Example 7.7

Evaluate the integral $\displaystyle\int_{t-2}^{\sin t} (x^2 - 2x + 3)\,dx$.

$$\int_{t-2}^{\sin t} (x^2 - 2x + 3)\,dx = \left[\frac{x^3}{3} - x^2 + 3x\right]_{t-2}^{\sin t}$$

$$= \frac{\sin^3 t}{3} - \sin^2 t + 3\sin t - \frac{(t-2)^3}{3} + (t-2)^2 - 3(t-2)$$

$$= \frac{\sin^3 t}{3} - \sin^2 t + 3\sin t - \frac{t^3}{3} + 3t^2 - 11t + \frac{38}{3}.$$

This example shows that when such an integral is evaluated the answer involves the variable which is present in the limits of integration. Now if we want to find the derivative of this expression we can evaluate the integral, as we have done in the above example, and then differentiate the answer. However we can find the derivative without integrating first, as follows.

Suppose that we know the indefinite integral of $f(x)$, i.e., that we know a function $F(x)$ satisfying $F'(x) = f(x)$. We then have

$$G(t) = \int_{a(t)}^{b(t)} f(x)\, dx = \int_{a(t)}^{b(t)} F'(x)\, dx = F(b(t)) - F(a(t)).$$

We can therefore differentiate using the chain rule to obtain

$$G'(t) = F'(b(t))b'(t) - F'(a(t))a'(t) = f(b(t))b'(t) - f(a(t))a'(t).$$

Example 7.8

Find the derivative with respect to t of the function defined by

$$F(t) = \int_{t^2-3t}^{t^3+4t^2} \cos(x)\, dx.$$

In this example we have $f(x) = \cos x$, $a(t) = t^2 - 3t$, and $b(t) = t^3 + 4t^2$. The formula we obtained tells us that the derivative is

$$F'(t) = (3t^2 + 8t)\cos(t^3 + 4t^2) - (2t - 3)\cos(t^2 - 3t).$$

In this case we could have integrated $\cos x$ first, to give

$$\int_{t^2-3t}^{t^3+4t^2} \cos(x)\, dx = [\sin x]_{t^2-3t}^{t^3+4t^2} = \sin(t^3 + 4t^2) - \sin(t^2 - 3t).$$

Applying the chain rule for differentiation will give the same answer as we obtained above using the differentiation formula.

Example 7.9

Find the derivative with respect to t of the function defined by

$$G(t) = \int_{t^2-3t}^{t^3+4t^2} \cos\left(x^2\right) dx.$$

In this example we cannot find the indefinite integral, but we can still use the formula for the derivative. This tells us that the derivative is

$$G'(t) = (3t^2 + 8t)\cos\left((t^3 + 4t^2)^2\right) - (2t - 3)\cos\left((t^2 - 3t)^2\right).$$

7.5 Infinite Integrals

So far we have considered definite integrals of the form $\int_a^b f(x)\,dx$, where a and b are real numbers, and for a non-negative function the integral corresponds to the area between the graph and the x-axis. In effect we are considering a function whose domain is limited to lie between a and b. But many functions have as their domain the set of all real numbers, or the set of all positive real numbers, or some other unbounded set. In this section we shall consider how to interpret the idea of the area of the region between such a graph and the x-axis. This leads to the idea of an **infinite integral**. Suppose we have a function $f(x)$ whose domain is the set of all real x satisfying $x \geq a$. We want to give a meaning to $\int_a^\infty f(x)\,dx$.

Imagine that we are going to paint such a region and that we want to know whether we can do it with a finite amount of paint. What we can do is to start at $x = a$ and paint the region up as far as $x = t$. We can then measure the amount of paint used. This will depend on t, and so can be expressed as a function $F(t)$. Using the ideas of limits from Chapter 2 we can then investigate $\lim_{t\to\infty} F(t)$. If this is finite it would appear that we can paint the complete region using a finite amount of paint. This somewhat far-fetched analogy motivates the definition.

Definition 7.10

If $f(x)$ is a function continuous for $x \geq a$ we define the infinite integral of f by

$$\int_a^\infty f(x)\,dx = \lim_{t\to\infty} \int_a^t f(x)\,dx.$$

If this limit exists and is finite we say that the infinite integral **converges**. Otherwise the infinite integral **diverges**.

We can similarly define

$$\int_{-\infty}^b f(x)\,dx = \lim_{t\to-\infty} \int_t^b f(x)\,dx.$$

Example 7.11

Investigate the convergence or otherwise of the infinite integral $\int_1^\infty x^\alpha\,dx$.

We will begin with two numerical cases by way of illustration.

Firstly let $\alpha = -2$. We then have

$$\int_1^t \frac{1}{x^2}\,dx = \left[-\frac{1}{x}\right]_1^t = 1 - \frac{1}{t} \to 1 \quad \text{as} \quad t \to \infty.$$

The integral therefore converges and we can write

$$\int_1^\infty \frac{1}{x^2}\,dx = 1.$$

Now consider the case $\alpha = -1$. We now have

$$\int_1^t \frac{1}{x}\,dx = [\ln x]_1^t = \ln t \to \infty \quad \text{as} \quad t \to \infty.$$

Therefore the integral in this case diverges.

In general suppose that $\alpha \neq -1$. Then

$$\int_1^t x^\alpha = \left[\frac{x^{\alpha+1}}{\alpha+1}\right]_1^t = \frac{t^{\alpha+1}}{\alpha+1} - \frac{1}{\alpha+1}.$$

Now if $\alpha + 1 > 0$ the right-hand expression tends to infinity, and so the corresponding infinite integral diverges. If $\alpha + 1 < 0$ then

$$\lim_{t\to\infty} \frac{t^{\alpha+1}}{\alpha+1} - \frac{1}{\alpha+1} = -\frac{1}{\alpha+1}.$$

Therefore the integral converges, and we can write

$$\int_1^\infty x^\alpha = -\frac{1}{\alpha+1}.$$

This is verified with the case $\alpha = -2$ we considered above.

Example 7.12

Show that the infinite integral $\int_0^\infty e^{-x}\,dx$ converges, and find its value.

Evaluating the integral over the finite interval $0 \leq x \leq t$ gives

$$\int_0^t e^{-x}\,dx = \left[-e^{-x}\right]_0^t = 1 - e^{-t} \to 1 \quad \text{as} \quad t \to \infty.$$

Therefore the integral converges and its value is 1.

Example 7.13

Show that the integral $\displaystyle\int_1^\infty e^{-x^2}\,dx$ converges.

In this case, unlike Example 7.12, we cannot evaluate the indefinite integral explicitly and so we cannot investigate the required limit directly. What we do is to compare this integral with one which we already know to converge. To return to the area interpretation, if we can show that the region between the graph of e^{-x^2} and the x-axis is smaller than one for which we already know that the area is finite then it too must enclose a finite area. The argument proceeds as follows in this case.

For all $x \geq 1$ we know that $0 \leq e^{-x^2} \leq e^{-x}$.

We therefore have, using the result of Example 7.12,

$$\int_0^t e^{-x^2}\,dx \leq \int_0^t e^{-x}\,dx \leq 1.$$

The value of the left-hand integral increases as t increases, because the integrand is positive for all x, and so it tends to a finite limit as $t \to \infty$. Therefore the integral converges.

Because we cannot evaluate the indefinite integral we do not know the value of the infinite integral. What we have showed is that

$$\int_0^\infty e^{-x^2}\,dx \leq 1.$$

The procedure used in this example generalises, as in the following theorem.

Theorem 7.14 (Comparison Test for Infinite Integrals)

Suppose that $f(x)$ and $g(x)$ are continuous, and that $0 \leq g(x) \leq f(x)$, for all $x \geq a$. Then if the infinite integral $\displaystyle\int_a^\infty f(x)\,dx$ converges, so does the infinite integral $\displaystyle\int_a^\infty g(x)\,dx$, and

$$\int_a^\infty g(x)\,dx \leq \int_a^\infty f(x)\,dx.$$

Proof

An analytical proof is outside the scope of this book. We shall give an intuitive geometrical explanation, which generalises the argument in Example 7.13.

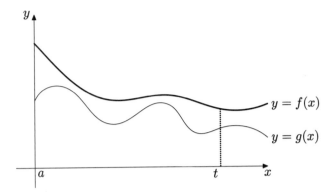

Figure 7.5   Integral comparison test

We can see from Figure 7.5 that the area between the graph of $f(x)$ and the x-axis is greater than the area between the graph of $g(x)$ and the x-axis. Therefore

$$\int_a^t g(x)\,dx \le \int_a^t f(x)\,dx.$$

From the figure we can also see that these areas increase as t increases, and because the functions are both non-negative it follows that the corresponding integrals also increase as t increases. If we denote the value of the convergent integral $\int_a^\infty f(x)\,dx$ by K we deduce that

$$\int_a^t g(x)\,dx \le K$$

for all $t \ge a$. So $\int_a^t g(x)\,dx$ is an increasing function of t which is bounded above. It therefore has a limit $H \le K$. (This last result would be proved in a course on Real Analysis, for example in Howie Chapter 3).   □

So far we have restricted attention to non-negative functions. Definition 7.10 does not require this, and in the next example we consider a function taking both signs.

Example 7.15

Show that the integral $\int_{2\pi}^\infty \frac{\sin x}{x^2}\,dx$ converges.

From Example 7.11 (with $\alpha = -2$) we know that the area between the graph of $1/x^2$ and the x-axis is finite. We also know that

$$-\frac{1}{x^2} \le \frac{\sin x}{x^2} \le \frac{1}{x^2}.$$

We can therefore see, in Figure 7.6, that the areas contained by those parts of the graph of $\dfrac{\sin x}{x^2}$ above the x-axis will be finite in total. The same will be true below the x-axis, so that the total area between the graph of $\dfrac{\sin x}{x^2}$ and the x-axis will be finite. Now the integral is found by subtracting the total area below the axis from the total area above the axis, and this will therefore be finite. In other words the integral $\displaystyle\int_{2\pi}^{\infty} \frac{\sin x}{x^2}\, dx$ will converge.

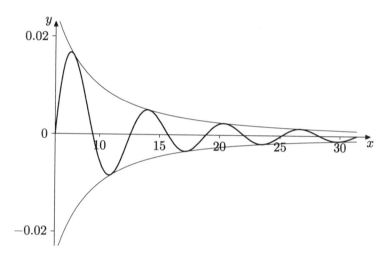

Figure 7.6   Graph for Example 7.15

This is an example of a generalisation of Theorem 7.14 which we state here without proof.

Theorem 7.16

Suppose that $f(x)$ and $g(x)$ are continuous, and that $|g(x)| \le f(x)$, for all $x \ge a$. Then if the infinite integral $\displaystyle\int_{a}^{\infty} f(x)\, dx$ converges, so does the infinite integral $\displaystyle\int_{a}^{\infty} g(x)\, dx$, and

$$\left| \int_{a}^{\infty} g(x)\, dx \right| \le \int_{a}^{\infty} |g(x)|\, dx \le \int_{a}^{\infty} f(x)\, dx.$$

7.6 Improper Integrals

In this section we consider integrals of the form $\int_a^b f(x)\,dx$, where the function f is undefined, or has a discontinuity, or is unbounded, at some point of the interval $a \le x \le b$. If this occurs inside the interval at some point c we can split the integral into two by using the rule of integration

$$\int_a^b f(x)\,dx = \int_a^c f(x)\,dx + \int_c^b f(x)\,dx.$$

This means that we can restrict attention to situations where the point of discontinuity etc. occurs at an end-point of the interval. We shall in fact discuss only cases where this happens at the lower end-point, the case of the upper end-point being exactly similar. So by way of example we can consider the integral $\int_0^1 \frac{1}{x^2}\,dx$, where the integrand is undefined at the end-point $x = 0$. Such an integral is called an **improper integral**, and we need to investigate questions of convergence. As with infinite integrals we are asking whether we can sensibly define the area of the region between the graph and the x-axis when the region is unbounded.

Definition 7.17

Let $f(x)$ be a function continuous for $a < x \le b$. Then the improper integral $\int_a^b f(x)\,dx$ is said to converge if the integral $\int_c^b f(x)\,dx$ has a limit as $c \to a^+$. We then define

$$\int_a^b f(x)\,dx = \lim_{c \to a^+} \int_c^b f(x)\,dx.$$

Example 7.18

Investigate convergence of the improper integral $\int_0^1 \frac{1}{x^2}\,dx$.

Using Definition 7.17, we have

$$\int_c^1 \frac{1}{x^2}\,dx = \left[-\frac{1}{x} \right]_c^1 = \frac{1}{c} - 1 \to \infty \quad \text{as} \ \ c \to 0^+.$$

Therefore the integral $\int_0^1 \frac{1}{x^2}\,dx$ does not converge.

Example 7.19

Investigate convergence of the improper integral $\int_0^1 \frac{1}{\sqrt{x}}\, dx$.

Using Definition 7.17 gives

$$\int_c^1 \frac{1}{\sqrt{x}}\, dx = \left[2\sqrt{x}\right]_c^1 = 2 - 2\sqrt{c} \to 2 \quad \text{as} \quad c \to 0^+.$$

Therefore the integral converges, and $\int_0^1 \frac{1}{\sqrt{x}}\, dx = 2$.

There are comparison tests for convergence of improper integrals like Theorems 7.14 and 7.16. Their discussion is left as an exercise for the reader. The following example is an application of the analogue of Theorem 7.16 for improper integrals.

Example 7.20

Show that the improper integral $\int_0^1 \frac{\sin(3x^2 + 2x - 4)}{\sqrt{x} + \sqrt[3]{x}}\, dx$ converges.

To apply a comparison test we need to compare the integrand with a function we know how to integrate. The numerator takes both positive and negative values in the interval $0 \le x \le 1$, but for all x, $|\sin(2x^2 + 2x - 4)| \le 1$. For the denominator, we note that in this interval $\sqrt[3]{x} \ge \sqrt{x}$, and so $\sqrt{x} + \sqrt[3]{x} \ge 2\sqrt{x}$. We therefore have, for all x in the interval,

$$\left| \frac{\sin(3x^2 + 2x - 4)}{\sqrt{x} + \sqrt[3]{x}} \right| \le \frac{1}{2\sqrt{x}}.$$

From Example 7.19 we know that the integral of the right-hand expression converges, and so by comparison the integral

$$\int_0^1 \frac{\sin(3x^2 + 2x - 4)}{\sqrt{x} + \sqrt[3]{x}}\, dx$$

converges. Using the result of Example 7.19 then tells us that

$$\left| \int_0^1 \frac{\sin(3x^2 + 2x - 4)}{\sqrt{x} + \sqrt[3]{x}}\, dx \right| \le \int_0^1 \left| \frac{\sin(3x^2 + 2x - 4)}{\sqrt{x} + \sqrt[3]{x}} \right|\, dx \le \int_0^1 \frac{1}{2\sqrt{x}}\, dx \le 1.$$

EXERCISES

7.1. Find the values of the following definite integrals by interpreting them as areas.

(a) $\displaystyle\int_{-3}^{2} (3x - 2)\, dx;$ (b) $\displaystyle\int_{-2}^{2} \sqrt{4 - x^2}\, dx;$

(c) $\displaystyle\int_{-1}^{3} |x - 2|\, dx;$ (d) $\displaystyle\int_{-4}^{3} \big| |x - 2| - |x + 2| \big|\, dx;$

(e) $\displaystyle\int_{-2}^{2} \sinh x\, dx;$ (f) $\displaystyle\int_{0}^{\pi} \sin 2x\, dx.$

7.2. Evaluate the following indefinite integrals.

(a) $\displaystyle\int (3x^2 + 4x - 2)\, dx;$ (b) $\displaystyle\int \sqrt{3x - 1}\, dx;$

(c) $\displaystyle\int (x^3 - 2)^2\, dx;$ (d) $\displaystyle\int \left(\sqrt{x} + x^2\right)^3\, dx;$

(e) $\displaystyle\int \left(\frac{1}{x^2} + \frac{4}{x^3}\right) dx;$ (f) $\displaystyle\int \frac{1}{\sqrt[3]{x}}\, dx;$

(g) $\displaystyle\int \frac{1}{\sqrt{2x + 3}}\, dx;$ (h) $\displaystyle\int e^{2x+3}\, dx;$

(i) $\displaystyle\int 2^{-x}\, dx;$ (j) $\displaystyle\int \cosh 3x\, dx;$

(k) $\displaystyle\int \sin x \cos x\, dx;$ (l) $\displaystyle\int \sec^2 x\, dx;$

(m) $\displaystyle\int \cos^2 x\, dx;$ (n) $\displaystyle\int \sin 2x \sin 5x\, dx.$

7.3. Find the derivative with respect to t of the function defined by

$$F(t) = \int_{t^2}^{t^3} e^x\, dx,$$

(a) by integrating first and then differentiating,

(b) by using the formula for the derivative.

7.4. Find the derivative with respect to t of the function defined by

$$G(t) = \int_{\sin t}^{\cos t} e^{(x^2)}\, dx.$$

7.5. For each of the following infinite integrals, determine whether or not it converges. Find the values of those that do converge.

(a) $\displaystyle\int_1^\infty \frac{1}{(2x+1)^2}\,dx;$ (b) $\displaystyle\int_1^\infty \frac{1}{\sqrt{x}}\,dx;$

(c) $\displaystyle\int_1^\infty \frac{1}{(3x+2)^{\frac{2}{3}}}\,dx;$ (d) $\displaystyle\int_0^\infty e^{2-3x}\,dx.$

7.6. Use a comparison test to show that the following infinite integrals converge.

(a) $\displaystyle\int_1^\infty \frac{e^{-x}}{\sqrt{x}}\,dx;$ (b) $\displaystyle\int_0^\infty \frac{e^{\sin x}}{1+x^2}\,dx;$

(c) $\displaystyle\int_3^\infty \frac{\sqrt{x^2-2x-2}}{x^3+x+4}\,dx;$ (d) $\displaystyle\int_0^\infty e^{-(x^3+x-3)}\,dx.$

7.7. Deduce the following result from Theorem 7.14.

Suppose that $f(x)$ and $g(x)$ are continuous, and that for all $x \geq a$, $0 \leq g(x) \leq f(x)$. Then if the infinite integral $\displaystyle\int_a^\infty g(x)\,dx$ does not converge, neither does the infinite integral $\displaystyle\int_a^\infty f(x)\,dx.$

7.8. For each of the following improper integrals, determine whether or not it converges. Find the values of those that do converge.

(a) $\displaystyle\int_2^3 \frac{1}{\sqrt[3]{x-2}}\,dx;$ (b) $\displaystyle\int_{-1}^3 \frac{1}{x+1}\,dx;$

(c) $\displaystyle\int_0^1 \frac{x+1}{\sqrt{x}}\,dx;$ (d) $\displaystyle\int_0^{\frac{\pi}{4}} \operatorname{cosec}^2 x\,dx.$

7.9. Determine the values of p for which the improper integral $\displaystyle\int_0^1 x^{-p}\,dx$ converges.

7.10. Formulate Theorems for improper integrals analogous to the comparison tests (Theorem 7.14 and Theorem 7.16) for infinite integrals. Give proofs in the same style as the proofs for infinite integrals in this chapter.

7.11. Use a comparison test to show that the following improper integrals converge.

(a) $\displaystyle\int_0^1 \frac{\sin x}{\sqrt{x}}\,dx;$ (b) $\displaystyle\int_{-2}^2 \frac{e^{-x}}{\sqrt{x+2}}\,dx.$

8

Integration by Parts

8.1 The Basic Technique

We are considering integration as the reverse of differentiation, and we should therefore expect that rules of differentiation should relate to techniques of integration. The technique of integration by parts discussed in this chapter is a consequence of the product rule for differentiation. That rule tells us that

$$\frac{d}{dx}\left(U(x)V(x)\right) = U(x)\frac{dV}{dx} + V(x)\frac{dU}{dx}.$$

If we integrate both sides with respect to x we obtain

$$U(x)V(x) = \int U(x)\frac{dV}{dx}\,dx + \int V(x)\frac{dU}{dx}\,dx,$$

which can be rearranged in the form

$$\int U(x)\frac{dV}{dx}\,dx = U(x)V(x) - \int V(x)\frac{dU}{dx}\,dx.$$

This equation therefore gives us a procedure for evaluating an integral of the form $\int f(x)g(x)\,dx$. We have to decide which of f and g to identify with U.

If we choose $f = U$ this is usually because $\dfrac{dU}{dx}$ is a simpler expression than U.

We then have to identify g with $\dfrac{dV}{dx}$, and we have to be able to find V, i.e., we have to be able to integrate $g(x)$.

Example 8.1

Evaluate $\int x \cos x \, dx$.

The integrand is a product, and so we have an expression for which integration by parts is a possible technique. So we have to decide which of the two components (x or $\cos x$) to identify with U. It is clear that if we choose $U = x$ then $\dfrac{dU}{dx} = 1$, which is simpler, whereas if we choose $\dfrac{dV}{dx} = x$ then $V = \dfrac{x^2}{2}$, which will give a more complicated integral. So we choose $\dfrac{dV}{dx} = \cos x$, which gives $V = \sin x$. The formula for integration by parts therefore gives

$$\int x \cos x \, dx = x \sin x - \int \sin x . 1 \, dx = x \sin x + \cos x,$$

which can be checked by differentiation.

Example 8.2

Evaluate $\int \ln x \, dx$.

This is a slightly less obvious application of integration by parts, since the integrand does not appear explicitly as a product. We can use the device of writing $\ln x$ as $1. \ln x$, and then identifying $U = \ln x$ and $\dfrac{dV}{dx} = 1$. Integration by parts then gives

$$\int 1. \ln x \, dx = x \ln x - \int x . \frac{1}{x} \, dx = x \ln x - \int 1 \, dx = x \ln x - x,$$

which can be checked by differentiation.

Example 8.3

Evaluate $\int x^2 \sinh 2x \, dx$.

In this case we shall choose $U = x^2$, because this simplifies when we differentiate. Integration by parts then gives

$$\int x^2 \sinh 2x \, dx = x^2 \frac{\cosh 2x}{2} - \int 2x \frac{\cosh 2x}{2} \, dx = x^2 \frac{\cosh 2x}{2} - \int x \cosh 2x \, dx.$$

The integral on the right-hand side again needs to be evaluated by parts, giving

$$\int x \cosh 2x \, dx = x \frac{\sinh 2x}{2} - \int \frac{\sinh 2x}{2} \, dx = x \frac{\sinh 2x}{2} - \frac{\cosh 2x}{4}.$$

Combining the two steps then gives

$$\int x^2 \sinh x \, dx = x^2 \frac{\cosh 2x}{2} - x\frac{\sinh 2x}{2} + \frac{\cosh 2x}{4}.$$

Example 8.4

Evaluate $I = \int e^x \sin 2x \, dx$.

In this example it is not immediately clear which of the two components of the product to choose for U. In each case the derivative is no simpler, and we can integrate either expression. In fact we could choose either $U = e^x$ or $U = \sin 2x$ in this example. We shall choose the former, and leave the other choice as an exercise for the reader to carry out. As in Example 8.3 we find that integrating by parts once does not leave us with a straightforward final integral, so we repeat the calculation. We shall find that we end up with the integral we started with, but in fact this then gives us an equation which we can solve to find I.

$$
\begin{aligned}
I &= \int e^x \sin 2x \, dx = e^x \left(-\frac{\cos 2x}{2} \right) + \int e^x \frac{\cos 2x}{2} \, dx \\
&= -e^x \left(\frac{\cos 2x}{2} \right) + e^x \frac{\sin 2x}{4} - \int e^x \frac{\sin 2x}{4} \, dx \\
&= \left[e^x \left(\frac{\sin 2x}{4} - \frac{\cos 2x}{2} \right) \right] - \frac{1}{4} I.
\end{aligned}
$$

We therefore have

$$I = \left[e^x \left(\frac{\sin 2x}{4} - \frac{\cos 2x}{2} \right) \right] - \frac{1}{4} I,$$

giving

$$\frac{5}{4} I = \left[e^x \left(\frac{\sin 2x}{4} - \frac{\cos 2x}{2} \right) \right],$$

and so finally

$$I = \frac{e^x (\sin 2x - 2 \cos 2x)}{5}.$$

8.2 Reduction Formulae

In Example 8.3 we evaluated $\int x^2 \sinh x \, dx$, integrating by parts twice. If we were faced with $\int x^6 \sinh x \, dx$ we would have to integrate by parts six times.

Not only would this be tedious, but we would find that the structure of successive steps would be similar. In this section we develop an approach which involves formulating a generic step, for which the result can easily be applied the requisite number of times, sometimes using a spreadsheet or a simple computer program. Such a generic step is called a reduction formula, because the procedure involves an integer parameter which is reduced in value at each stage of the process, until we reach a stage where the resulting integral is easily evaluated. We can find reduction formulae for both indefinite and definite integrals, as in the next example.

Example 8.5

Find reduction formulae for $I_n = \int x^n e^x \, dx$ and $J_n = \int_0^1 x^n e^x \, dx$.

The notation I_n indicates that the integral involves the integer parameter n. The answer depends upon the value of n, as for example with

$$\int_0^1 x^n \, dx = \left[\frac{x^{n+1}}{n+1}\right]_0^1 = \frac{1}{n+1} \quad (n \neq -1),$$

where we can see explicitly that the answer involves n.

In this example we have, in the case of the indefinite integral,

$$I_n = \int x^n e^x \, dx = x^n e^x - \int n x^{n-1} e^x \, dx.$$

We can see that the final integral is obtained from I_n by replacing n by $n-1$ and multiplying by n. This is expressed in the reduction formula

$$I_n = x^n e^x - n I_{n-1}.$$

In the case of the definite integral the calculations are the same, and so we obtain

$$J_n = \int_0^1 x^n e^x \, dx = [x^n e^x]_0^1 - \int_0^1 n x^{n-1} e^x \, dx,$$

which gives the reduction formula

$$J_n = e - n J_{n-1}.$$

We can use this formula to work out, for example, the value of $J_7 = \int_0^1 x^7 e^x \, dx$.

We first apply the reduction formula with $n = 7$, giving $J_7 = e - J_6$. We now apply the same reduction formula, but with $n = 6$, and so

$$J_7 = e - 7J_6 = e - 7(e - 6J_5) = -6e + 7.6J_5 = -6e + 42J_5.$$

This process continues until we reach J_0, which we can evaluate explicitly, because

$$J_0 = \int_0^1 x^0 e^x \, dx = e - 1.$$

The complete process using the reduction formula for values of n from 7 down to 1 is as follows.

$$
\begin{aligned}
J_7 &= e - 7J_6 \\
&= e - 7(e - 6J_5) = -6e + 42J_5 \\
&= -6e + 42(e - 5J_4) = 36e - 210J_4 \\
&= 36e - 210(e - 4J_3) = -174e + 840J_3 \\
&= -174e + 840(e - 3J_2) = 666e - 2520J_2 \\
&= 666e - 2520(e - 2J_1) = -1854e + 5040J_1 \\
&= -1854e + 5040(e - 1.J_0) = -1854e + 5040(e - (e - 1)) \\
&= -1854e + 5040.
\end{aligned}
$$

Example 8.6

Find a reduction formula for $I_n = \int_0^\pi x^n \sin x \, dx$.

We shall see that we need to integrate by parts twice in this case, in order to return to an integral containing $\sin x$.

$$
\begin{aligned}
I_n &= \int_0^\pi x^n \sin x \, dx = [x^n(-\cos x)]_0^\pi + \int_0^\pi nx^{n-1} \cos x \, dx \\
&= \pi^n + \int_0^\pi nx^{n-1} \cos x \, dx \\
&= \pi^n + [nx^{n-1} \sin x]_0^\pi - \int_0^\pi n(n-1)x^{n-2} \sin x \, dx \\
&= \pi^n - n(n-1)I_{n-2}.
\end{aligned}
$$

So the reduction formula is

$$I_n = \pi^n - n(n-1)I_{n-2}.$$

In Example 8.5 the integer parameter n was reduced by 1 at each stage. In this example n is reduced by 2 each time. So if we begin with an odd value of n we shall finish by needing to calculate I_1, and if we start with an even value of n we shall need the value of I_0. In both cases these are integrals which are easy to evaluate explicitly.

8.3 Integration using MAPLE

If we simply want to know the value of an indefinite or a definite integral MAPLE will provide the answer. So in Example 8.1 if we use the command

```
int(x*cos(x),x);
```

we will immediately obtain the output $\cos x + x \sin x$, which agrees with our calculations. Similarly for Example 8.3 the command

```
int(x^2*sinh(2*x),x);
```

gives an output which agrees with our calculations.

MAPLE can also evaluate definite integrals, and the command

```
int(x^3*sin(x),x=0..Pi);
```

yields $-6\pi + \pi^3$. To evaluate this by hand would require integration by parts three times.

MAPLE can do much more than simply provide the answers. It has a library package which will demonstrate the various steps in a process such as integration by parts. To load the package in question we use the command `with(student);` and we then see a list of the additional commands this package contains. To see the process of integration by parts at work we tell MAPLE which function to use as U.

Example 8.7

Work through Example 8.3 using MAPLE as explained above.

We use commands as follows.

```
with(student);
A:=Int(x^2*sinh(2*x),x);
```

The output is $A := \int x^2 \sinh(2x)\, dx$. We see that the command gives the integral a name, so it means we do not have to re-type it every time we want to use it.

```
intparts(A,x^2);
```

This command tells MAPLE to take $U = x^2$. The output agrees precisely with the first step of our calculations. We need to integrate by parts again, and so the command

```
intparts(%,x);
```

takes the previous result (denoted in MAPLE by the percentage symbol) and evaluates the integral with $U = x$, the output agreeing with the second stage of our calculations. Finally we would simply evaluate the final straightforward integral.

Now in some cases, as in Example 8.4, it is not clear what to take for U. With MAPLE we can try both functions and see what the result is. In Example 8.1, if we take $U = \cos x$ instead of $U = x$, the commands

```
B:=Int(x*cos(x),x);
intparts(B,cos(x));
```

give the output

$$\frac{1}{2} \cos(x)x^2 - \int -\frac{1}{2} \sin(x)\, dx.$$

This is clearly a more complicated integral than the one we started with, so that $U = \cos(x)$ is an inappropriate choice.

Example 8.8

Use MAPLE to find a reduction formula for the integral $J(n)$ in Example 8.5.

The first command we need is `restart; with(student);` which clears the memory and then loads the relevant library routines. Next we label the integral, and the command includes the fact that it is a function of the integer variable n, as explained in Example 8.5.

```
J:=n->Int(x^n*exp(x),x=0..1);
```

We then instruct MAPLE to integrate by parts, taking $U = x^n$, and tell it to simplify the result.

```
intparts(J(n),x^n);
simplify(%);
```

The output is $\quad e - n \int_0^1 x^{(n-1)} e^x\, dx$, i.e., $\quad e - nJ(n-1)$.

To work out $J(7)$ using the reduction formula MAPLE needs to know the value of $J(0)$, as the procedure in Example 8.5 explains. So we first need the command `value(J(0));` which tells us that $J(0) = e - 1$, and then finally `value(J(7));` gives the answer $-1854e + 5040$.

8.4 The Gamma Function

In this section we consider some elementary properties of the Gamma function. This is a function encountered in applied mathematics and in statistics. It brings together integration by parts and infinite and improper integrals discussed in Sections 7.5 and 7.6.

Definition 8.9

The Gamma function is a function of the variable x defined by the integral

$$\Gamma(x) = \int_0^\infty t^{x-1} e^{-t} \, dt.$$

This is an infinite integral, and so we need to discuss convergence, as in Section 7.5.

If $x - 1 < 0$ then the integrand contains a negative power of t, which is undefined at $t = 0$. In this case we therefore have an improper integral, and convergence at $t = 0$ must be investigated, as in Section 7.6.

We shall investigate these two cases of convergence separately by splitting the integral at $t = 1$, letting

$$I_1 = \int_0^1 t^{x-1} e^{-t} \, dt, \quad I_2 = \int_1^\infty t^{x-1} e^{-t} \, dt.$$

In both cases we use comparison to discuss convergence.

Consider I_1. Since $0 \le t \le 1$ we have $e^{-1} \le e^{-t} \le 1$, so

$$t^{x-1} e^{-1} \le t^{x-1} e^{-t} \le t^{x-1}.$$

If $x > 0$ then for $0 < h < 1$ we have

$$\int_h^1 t^{x-1} \, dt = \left[\frac{t^x}{x} \right]_h^1 = \frac{1}{x}(1 - h^x) \to \frac{1}{x} \quad \text{as} \ h \to 0.$$

Hence $\int_0^1 t^{x-1} \, dt$ converges, and so $\int_0^1 t^{x-1} e^{-t} \, dt$ converges by comparison.

Now if $x < 0$ then for $0 < h < 1$ we have

$$\int_h^1 e^{-1} t^{x-1} \, dt = e^{-1} \left[\frac{t^x}{x} \right]_h^1 = \frac{e^{-1}}{x}(1 - h^x) \to \infty \quad \text{as} \ h \to 0.$$

Hence $\int_0^1 e^{-1} t^{x-1} \, dt$ diverges, and so $\int_0^1 t^{x-1} e^{-t} \, dt$ diverges by comparison.

Now we consider I_2. Let n denote the first integer greater than x. We therefore have, using the method of Example 6.14

$$t^{x-1} e^{-t} < t^{n-1} e^{-t} < t^{n-1} \frac{(n+1)!}{t^{n+1}} = \frac{(n+1)!}{t^2}.$$

In Example 7.11 we saw that $\int_1^\infty \frac{1}{t^2} \, dt$ converges. Therefore by comparison (noting that $(n+1)!$ is a constant), $\int_1^\infty t^{x-1} e^{-t} \, dt$ converges.

We would like to integrate by parts, but we have an infinite (and possibly improper) integral, so this needs careful consideration. Suppose that $x > 1$, so that the integral is not improper at $x = 0$. We then integrate by parts as follows.

$$\int_0^k t^x e^{-t}\, dt = \left[-t^x e^{-t}\right]_0^k + \int_0^k xt^{x-1}e^{-t}\, dt.$$

Letting $k \to \infty$ then gives

$$\int_0^\infty t^x e^{-t}\, dt = x \int_0^\infty t^{x-1}e^{-t}\, dt, \quad \text{i.e.,} \quad \Gamma(x+1) = x\Gamma(x).$$

This argument will generalise to show that we can integrate convergent infinite integrals by parts. We can deal with the case $0 < x \le 1$, when the integral is improper, in a similar fashion, using the limit definition of an improper integral as we did for an infinite integral above. So we can say that $\Gamma(x+1) = x\Gamma(x)$ for $x > 0$.

This looks like a reduction formula (see Section 8.2), and if we let $x = n$, a positive integer, we can see that

$$\begin{aligned}
\Gamma(n+1) &= n\Gamma(n) = n(n-1)\Gamma(n-1) \\
&= n(n-1)(n-2)\Gamma(n-2) \\
&= n(n-1)(n-2)\ldots 2.1.\Gamma(1).
\end{aligned}$$

Now $\Gamma(1) = \int_1^\infty t^0 e^{-t}\, dt = \left[-e^{-t}\right]_0^\infty = 1$, so $\Gamma(n+1) = n!$ The Gamma function can therefore be seen as a generalisation of the factorial function for non-integer values.

One interesting result for a non-integer value is $\Gamma\left(\frac{1}{2}\right) = \sqrt{\pi}$. MAPLE knows this fact about the Gamma function, as the command GAMMA(1/2); will show. This and further information about the Gamma function can be found on various mathematical websites, for example the following two.

http://mathworld.wolfram.com
http://numbers.computation.free.fr

8.5 A Strange Example

We integrate $\tan x$ by parts, using $U = \dfrac{1}{\cos x}, \dfrac{dV}{dx} = \sin x$.

$$\int \tan x\, dx = \int \frac{\sin x}{\cos x}\, dx$$

$$= \frac{-\cos x}{\cos x} + \int -\cos x \cdot \frac{-\sin x}{\cos^2 x} \, dx$$

$$= -1 + \int \tan x \, dx.$$

Cancelling $\int \tan x \, dx$ therefore gives $0 = -1$.

So how do we explain this apparent paradox? Is it because of the constant of integration? Well if we were to include that we would get

$$\int \tan x \, dx + C = -1 + \int \tan x \, dx + C,$$

suggesting that $C = -1 + C$, which again is wrong. If the constant of integration is arbitrary perhaps we should not have the same constant C on both sides. We would then obtain

$$\int \tan x \, dx + C = -1 + \int \tan x \, dx + D.$$

This would imply that $C = -1 + D$, but if C and D are arbitrary constants why should they be related? What we need to do is to interpret C and D as representing the set of all possible constants, so that if C and D represent all possible real numbers then the set of all numbers of the form $-1 + D$ is also the set of all real numbers, the same as C. From another perspective it means that we have to think carefully about what an indefinite integral is. This example suggests that an indefinite integral is not a function, but a set of functions, and so the two sets of functions on either side of the equation

$$\int \tan x \, dx + C = -1 + \int \tan x \, dx + C$$

are identical, and there is no question of cancelling $\int \tan x \, dx$.

When we introduced indefinite integrals in Section 7.2 we used the indefinite article, and described a function $F(x)$ whose derivative is $f(x)$ as **an** indefinite integral (and not **the** indefinite integral). This is normal usage, and it would be very complicated to develop the procedures of integration in terms of the language of sets of functions. We shall not therefore change our approach, but simply be aware that occasionally we may need to think more precisely about the definition of an indefinite integral if we encounter an apparent paradox.

EXERCISES

8.1. Evaluate the following indefinite integrals

(a) $\int x \sinh x \, dx$;

(b) $\int x^2 e^x \, dx$;

(c) $\int x^2 \cosh 3x \, dx$;

(d) $\int x (\ln x)^2 \, dx$;

(e) $\int x^2 \cos x \, dx$;

(f) $\int x \tan^{-1} x \, dx$;

(g) $\int x \cos^2 x \, dx$;

(h) $\int \sqrt{x} \ln \left(\sqrt{x} \right) \, dx$;

(i) $\int (x + 1) e^x \ln x \, dx$;

(j) $\int \cos^{-1} x \, dx$.

8.2. Use integration by parts to evaluate $\displaystyle\int e^{ax} \cos bx \, dx$.

8.3. Evaluate the indefinite integral $\displaystyle\int \sec^3 x \, dx$, by writing

$\sec^3 x = \sec x \sec^2 x$ and using integration by parts.

8.4. Find a reduction formula for $\displaystyle\int_1^e x (\ln x)^n \, dx$.

Hence evaluate $\displaystyle\int_1^e x (\ln x)^4 \, dx$.

8.5. Let $I_n = \displaystyle\int_0^{\frac{\pi}{2}} \cos^n x \, dx$.

Write $\cos^n x = \cos x \cos^{n-1} x$ and show that $I_n = \dfrac{n-1}{n} I_{n-2}$.

Hence evaluate $\displaystyle\int_0^{\frac{\pi}{2}} \cos^8 x \, dx$.

8.6. Use MAPLE to set up a reduction formula for $\displaystyle\int_0^1 x^n e^{2x} \, dx$, and use
it to find the value of the integral when $n = 7$.

8.7. Assuming that $\Gamma \left(\frac{1}{2} \right) = \sqrt{\pi}$, find the values of $\Gamma \left(\frac{3}{2} \right)$ and $\Gamma \left(\frac{7}{2} \right)$.

9
Integration by Substitution

The theoretical basis for integration by substitution is the chain rule for differentiation, which says that

$$\frac{d}{dx} f(g(x)) = f'(g(x)) g'(x).$$

Regarding integration as the reverse of differentiation leads us to integrate both sides of this equation to write

$$\int f'(g(x)) g'(x)\, dx = f(g(x)).$$

Given an integral of this form we can transform it by means of the substitution $u = g(x)$. We then have $\dfrac{du}{dx} = g'(x)$ and so

$$\int f'(g(x)) g'(x)\, dx = \int f'(u) \frac{du}{dx}\, dx = \int \frac{d}{dx} (f(u))\, dx = f(u) = f(g(x)),$$

where we have used the chain rule with the intermediate variable u, to recognise that $f'(u)\dfrac{du}{dx} = \dfrac{d}{dx}(f(u))$.

In fact we implement this process symbolically by rewriting $\dfrac{du}{dx} = g'(x)$ in the form $du = g'(x)\, dx$. The integration procedure then appears in the form

$$\int f'(g(x)) g'(x)\, dx = \int f'(u)\, du = f(u) = f(g(x)).$$

The underlying idea is that the substitution gives rise to a simpler integral involving the variable u. After having evaluated this integral we then replace

u in the answer by $g(x)$, so as to present the answer in terms of the original variable x.

This can all be made analytically rigorous. The details are beyond the scope of this book. In the remainder of this chapter we shall concentrate therefore on applying this technique of integration in a variety of circumstances.

9.1 Some Simple Substitutions

Example 9.1

Evaluate the indefinite integral $\displaystyle\int \frac{\cos x}{(1 + \sin x)^3}\, dx$.

The theory above looks straightforward, but if we are presented with an integral like $\displaystyle\int \frac{\cos x}{(1 + \sin x)^3}\, dx$, how are we to find an appropriate substitution which will transform the integral into a simpler one that we can recognise? We need to be able to discern what should play the role of f and what should play the role of g. We note that the general integral expression $\displaystyle\int f'\left(g(x)\right)g'(x)\, dx$ involves both g and its derivative, so what we need to look for is one part of the integrand which is the derivative of another part of the integrand. In this example we can see that the integrand involves $\sin x$ and also $\cos x$, which is the derivative of $\sin x$. This suggests using the substitution $u = \sin x$. We then have $du = \cos x\, dx$, and so the integral transforms as

$$\int \frac{\cos x}{(1 + \sin x)^3}\, dx = \int \frac{1}{(1 + u)^3}\, du.$$

The latter is an integral we should know how to do, but if not we can simplify it still further with a linear transformation $w = 1 + u$, giving $dw = du$, and therefore

$$\int \frac{\cos x}{(1 + \sin x)^3}\, dx = \int \frac{1}{(1 + u)^3}\, du = \int \frac{1}{w^3}\, dw = -\frac{1}{2w^2}.$$

The answer has to be given in terms of the original variable x, so we have

$$\int \frac{\cos x}{(1 + \sin x)^3}\, dx = -\frac{1}{2w^2} = -\frac{1}{2(1 + u)^2} = -\frac{1}{2(1 + \sin x)^2}.$$

The answer should be checked by differentiation (using the chain rule).

Example 9.2

Evaluate the indefinite integral $\displaystyle\int \frac{e^x}{1 + e^{2x}}\, dx$.

We first note that $e^{2x} = (e^x)^2$, so that the numerator is the derivative of part of the denominator. This suggests the substitution $u = e^x$, giving $du = e^x\, dx$. We therefore have

$$\int \frac{e^x}{1 + e^{2x}}\, dx = \int \frac{1}{1 + u^2}\, du = \tan^{-1} u = \tan^{-1}(e^x).$$

Example 9.3

Evaluate the definite integral $\displaystyle\int_1^{\sqrt{e}} \frac{\sin(\pi \ln x)}{x}\, dx$.

In this case when we look for one part of the integrand which is the derivative of another part, we see that the integrand contains $\ln x$ and its derivative $1/x$. We therefore use the substitution $u = \ln x$, giving $du = \frac{1}{x}\, dx$.

In the case of a definite integral we do not need to evaluate the corresponding indefinite integral in terms of the original variable, because the substitution transforms the interval of integration for x into an interval of integration for u. In order for this to work properly the substitution must have an inverse over the interval of integration, in other words the function used in the substitution has to be 1-1 over that interval. (Example 9.4 discusses this point.) This is the case here, $\ln x$ being 1-1 over its entire domain. Applying the substitution, we find that when $x = 1, u = \ln 1 = 0$, and when $x = \sqrt{e},\, u = \ln(\sqrt{e}) = \frac{1}{2}$. We can now transform the integral, giving

$$\int_{x=1}^{x=\sqrt{e}} \frac{\sin(\pi \ln x)}{x}\, dx = \int_{u=0}^{u=\frac{1}{2}} \sin(\pi u)\, du = \left[-\frac{\cos(\pi u)}{\pi} \right]_{u=0}^{u=\frac{1}{2}} = \frac{1}{\pi}.$$

In this example we have used an expanded notation for the limits of integration, to emphasise the point about transforming them at the same time as transforming the integrand. Normally we would just write the numbers without indicating the variables explicitly. However the notation is useful when integration of functions of more than one variable is studied in multivariable calculus.

Example 9.4

In this example we illustrate the comment in Example 9.3 about the substitution needing to have an inverse.

We shall try to evaluate $\displaystyle\int_0^{\pi} \cos^2 x\, dx$ using the substitution $u = \sin x$, which is not 1-1 over the interval $0 \leq x \leq \pi$. We then have $du = \cos x\, dx$, and using the identity $\cos^2 x + \sin^2 x = 1$ gives $\cos x = \sqrt{1 - u^2}$. Applying the substitution to the interval of integration, we find that when $x = 0, u = 0$, and when $x = \pi, u = 0$ also. So we end up with the integral $\displaystyle\int_0^0 \sqrt{1 - u^2}\, du$, which is zero, whereas the original integral is clearly positive. The substitution seems to give an erroneous result. If we were to use this substitution we would have to split the interval of integration into separate intervals on which the substitution would be 1-1. These could be the intervals $0 \le x \le \pi/2$ and $\pi/2 \le x \le \pi$ respectively.

Another aspect of this example is that the equation $\cos x = \sqrt{1 - u^2}$ is valid only for $0 \le x \le \pi/2$. For $\pi/2 \le x \le \pi$ we should have $\cos x = -\sqrt{1 - u^2}$. As in other situations we have to be very careful about the correct choice of square root. Taking this into account also implies that we should split the interval of integration.

9.2 Inverse Substitutions

In the previous section the substitutions used replaced part of the integrand by a single variable, as in Example 9.1 where we replaced $\sin x$ by u. In this section we consider substitutions which work in the reverse direction, so that x itself is replaced by an expression in another variable, for example we may substitute $x = \sin u$. We can regard this as equivalent to $u = \sin^{-1} x$, hence the term **inverse substitution**.

Example 9.5

Evaluate the indefinite integral $I = \displaystyle\int \frac{x^2}{\sqrt{x} + 2}\, dx$

It is the square root term which makes this integral slightly awkward, so we choose an inverse substitution to remove it, the obvious one being $x = u^2$. We then have $dx = 2u\, du$, and so

$$
\begin{aligned}
I &= \int \frac{u^4}{u + 2} \cdot 2u\, du = \int \frac{2u^5}{u + 2}\, du \\
&= \int \left(2u^4 - 4u^3 + 8u^2 - 16u + 32 - \frac{64}{u + 2} \right) du \\
&\quad \text{(using polynomial division)} \\
&= \frac{2u^5}{5} - u^4 + \frac{8u^3}{3} - 8u^2 + 32u - 64 \ln|u + 2| \\
&= \frac{2x^{\frac{5}{2}}}{5} - x^2 + \frac{8x^{\frac{3}{2}}}{3} - 8x + 32\sqrt{x} - 64 \ln|\sqrt{x} + 2|.
\end{aligned}
$$

Because $x = u^2$ is a simple inverse transformation it has a straightforward direct equivalent, namely $u = \sqrt{x}$. Applying this direct substitution however is slightly more awkward, because we obtain $du = \dfrac{dx}{\sqrt{x}}$, which still involves the square root, and for which the algebraic manipulation needed is slightly more involved.

Example 9.6

Evaluate the indefinite integral $I = \displaystyle\int \dfrac{1}{\sqrt{x}(\sqrt[3]{x} + 2)}\, dx$.

Here we need to look for a substitution which will eliminate both the square and the cube roots, and $x = u^6$ will achieve this. So we have $dx = 6u^5\, du$ and therefore

$$
\begin{aligned}
I &= \int \frac{1}{\sqrt{x}\left(\sqrt[3]{x} + 2\right)}\, dx = \int \frac{6u^5}{u^3(u^2 + 2)}\, du \\
&= 6\int \frac{u^2}{u^2 + 2}\, du = 6\int \left(1 - \frac{2}{u^2 + 2}\right) du \\
&= 6\left(u - \sqrt{2}\tan^{-1}\left(\frac{u}{\sqrt{2}}\right)\right) = 6\left(\sqrt[6]{x} - \sqrt{2}\tan^{-1}\left(\frac{\sqrt[6]{x}}{\sqrt{2}}\right)\right).
\end{aligned}
$$

In the remainder of this chapter we shall concentrate on some inverse substitutions which work for classes of integrals. They will be illustrated through particular examples, but discussed in a way which emphasises their general applicability.

9.3 Square Roots of Quadratics

These can generally be simplified by means of trigonometric or hyperbolic substitutions. A knowledge of trigonometric identities, and of the algebraic technique of completing the square, is a necessity therefore. The examples are chosen to involve only a single square root term, but the techniques are equally applicable to integrals of algebraic fractions where the numerator or the denominator is the square root of a quadratic.

Example 9.7

Evaluate the indefinite integral $I = \displaystyle\int \sqrt{4x^2 - 16x + 52}\, dx$.

We shall break this example down into a number of steps, each of which is important in evaluating integrals of this type.

STEP 1 We make the coefficient of x^2 equal to 1, so

$$I = 2 \int \sqrt{x^2 - 4x + 13} \, dx.$$

(Note that if the coefficient of x^2 is negative then we make its coefficient equal to -1 and then proceed with the following steps.)

STEP 2 Complete the square, giving

$$I = 2 \int \sqrt{(x-2)^2 + 9} \, dx.$$

STEP 3 Write the constant as a square number.

$$I = 2 \int \sqrt{(x-2)^2 + 3^2} \, dx.$$

STEP 4 Use a linear substitution to replace the variable square term with a single square, using $u = x - 2$ in this case.

$$I = 2 \int \sqrt{u^2 + 3^2} \, du.$$

This succession of steps will always transform the square root of a quadratic into a sum of squares, as in this example, or a difference of squares. The following steps are therefore applicable to integrals involving a term of the form $\sqrt{u^2 + a^2}$.

STEP 5 Make an appropriate trigonometric substitution. We need to find a substitution which makes use of a trigonometric identity reducing a sum of two squares to a single square term, which will then enable us to remove the square root. In this case the appropriate identity is $\tan^2 t + 1 = \sec^2 t$. However we have $\sqrt{u^2 + 3^2}$, and so we need to take the 3^2 into account, using $u = 3 \tan t$. We then have $du = 3 \sec^2 t \, dt$, and so the substitution gives

$$I = 2 \int \sqrt{3^2 \tan^2 t + 3^2} . 3 \sec^2 t \, dt = 2.3.3 \int \sec^3 t \, dt.$$

STEP 6 We now have to evaluate the trigonometric integral. This involves integration by parts, and one of the exercises in Chapter 8 describes how to do this. We find that

$$I = 18 \left(\tfrac{1}{2} \sec t \tan t + \tfrac{1}{2} \ln |\sec t + \tan t| \right).$$

STEP 7 We now have to express the result in terms of x, firstly by finding the trigonometric functions in terms of u. The substitution gives us one of them, because $\tan t = \dfrac{u}{3}$. To find $\sec t$ (or any other trigonometric function which might arise is integrals of this type) a helpful technique is to draw a right-angled triangle in which $\tan t = \dfrac{u}{3}$. Using Pythagoras' Theorem gives the third side of the triangle, and we can then read off any trigonometric ratio that we need.

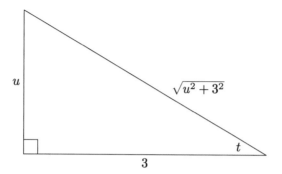

Figure 9.1   Trigonometric substitutions

We can now see from the diagram that $\sec t = \dfrac{\sqrt{u^2 + 3^2}}{3}$. So we can express the integral in terms of u as

$$I = 9\frac{\sqrt{u^2 + 3^2}}{3}\frac{u}{3} + 9\ln\left|\frac{\sqrt{u^2 + 3^2}}{3} + \frac{u}{3}\right|.$$

STEP 8 We replace u, using $u = x - 2$ from *STEP 4*, and undertake some algebraic simplification, to give the answer

$$I = \sqrt{(x-2)^2 + 3^2}(x-2) + 9\ln\left|\sqrt{(x-2)^2 + 3^2} + (x-2)\right| - 9\ln 3.$$

The $-9\ln 3$ term can be absorbed into the constant of integration, giving finally

$$I = (x-2)\sqrt{x^2 - 4x + 13} + 9\ln\left|\sqrt{x^2 - 4x + 13} + (x-2)\right|.$$

Example 9.8

Evaluate the indefinite integral $I = \displaystyle\int \frac{1}{\sqrt{x^2 - 2x + 6}}\,dx.$

We shall follow the steps as in Example 9.7. *STEP 1* is not needed here, but *STEP 2* and *STEP 3* have to be implemented. So completing the square gives

$$I = \int \frac{1}{\sqrt{(x-1)^2 + 5}}\,dx = \int \frac{1}{\sqrt{(x-1)^2 + \sqrt{5}^2}}\,dx.$$

Notice that we need to express the constant as a square even when it is not a square number like 9. This always serves as a reminder of the numerical factor needed in the substitution. We now perform *STEP 5* using $u = x - 1$, to give

$$I = \int \frac{1}{\sqrt{u^2 + \sqrt{5}^2}} \, du.$$

The trigonometric method used in the previous example would work here, but we shall show how to use the alternative of a hyperbolic substitution. The relevant hyperbolic identity is $\sinh^2 t + 1 = \cosh^2 t$, and so we use the substitution $u = \sqrt{5} \sinh t$, from which $du = \sqrt{5} \cosh t \, dt$. The integral then becomes

$$I = \int \frac{1}{\sqrt{u^2 + \sqrt{5}^2}} \, du = \int \frac{1}{\sqrt{5 \sinh^2 t + 5}} \sqrt{5} \cosh t \, dt.$$

This simplifies to give

$$I = \int \frac{\sqrt{5} \cosh t}{\sqrt{5} \cosh t} \, dt = \int 1 \, dt = t = \sinh^{-1} \left(\frac{u}{\sqrt{5}} \right) = \sinh^{-1} \left(\frac{x-1}{\sqrt{5}} \right).$$

In example 9.7 the answer involved a logarithmic function, and we might expect that here, since we are also integrating a function involving the square root of a quadratic. The apparent difference lies in the fact that inverse hyperbolic functions have equivalent logarithmic forms, as we saw in Example 1.40.

Example 9.9

Evaluate the indefinite integral $\displaystyle\int (x^2 - a^2)^{\frac{3}{2}} \, dx \quad (x^2 \geq a^2)$.

Although there is not a square root symbol, the integrand is of course the cube of a square root. This time we need an identity which changes a difference of two square terms into a single square term.

We can use either the trigonometric identity $\sec^2 t - 1 = \tan^2 t$, or the hyperbolic identity $\cosh^2 t - 1 = \sinh^2 t$. We shall use the latter, leaving the former as an exercise for the reader. Naturally they should give the same answer, or answers which are algebraically equivalent.

So we let $x = a \cosh t$, giving $dx = a \sinh t \, dt$. The integral then becomes

$$I = \int \left((a \cosh t)^2 - a^2 \right)^{\frac{3}{2}} a \sinh t \, dt = a^4 \int \sinh^4 t \, dt.$$

This integral can be evaluated either using hyperbolic double angle formulae,

$$\sinh 2t = 2 \sinh t \cosh t; \quad \cosh 2t = \cosh^2 t - \sinh^2 t = 2 \cosh^2 t - 1 = 2 \sinh^2 t + 1,$$

or using the exponential definition of hyperbolic functions. This is one advantage of using a hyperbolic substitution, for with trigonometric functions we do not have an exponential alternative. We shall demonstrate both methods, firstly using the hyperbolic identities. The calculation is quite long, and we have shown each step. The first steps of the calculation express powers of the hyperbolic functions in terms of individual hyperbolic functions which can be integrated. The latter steps of the calculation are concerned with expressing the answer back in terms of $\cosh t$ and $\sinh t$ in order to be able to obtain the result in terms of the original variable x. Readers will need to study each step of the calculation to see how the various identities have been used.

$$
\begin{aligned}
I &= a^4 \int \sinh^4 t \, dt = a^4 \int (\sinh^2 t)^2 \, dt \\
&= a^4 \int \left(\frac{\cosh 2t - 1}{2} \right)^2 \, dt \\
&= a^4 \int \left(\frac{\cosh^2 2t}{4} - \frac{\cosh 2t}{2} + \frac{1}{4} \right) \, dt \\
&= a^4 \int \left(\frac{\cosh 4t + 1}{8} - \frac{\cosh 2t}{2} + \frac{1}{4} \right) \, dt \\
&= a^4 \int \left(\frac{\cosh 4t}{8} - \frac{\cosh 2t}{2} + \frac{3}{8} \right) \, dt \\
&= a^4 \left(\frac{\sinh 4t}{32} - \frac{\sinh 2t}{4} + \frac{3t}{8} \right) \\
&= a^4 \left(\frac{\sinh 2t \cosh 2t}{16} - \frac{\sinh t \cosh t}{2} + \frac{3t}{8} \right) \\
&= a^4 \left(\frac{\sinh t \cosh t (2 \sinh^2 t + 1)}{8} - \frac{\sinh t \cosh t}{2} + \frac{3t}{8} \right) \\
&= a^4 \left(\frac{\sinh^3 t \cosh t}{4} - \frac{3 \sinh t \cosh t}{8} + \frac{3t}{8} \right) \\
&= a^4 \left(\frac{1}{4} \left(\frac{x^2}{a^2} - 1 \right)^{\frac{3}{2}} \frac{x}{a} - \frac{3}{8} \left(\frac{x^2}{a^2} - 1 \right)^{\frac{1}{2}} \frac{x}{a} + \frac{3}{8} \cosh^{-1} \left(\frac{x}{a} \right) \right) \\
&= \frac{x(x^2 - a^2)^{\frac{3}{2}}}{4} - \frac{3a^2 x(x^2 - a^2)^{\frac{1}{2}}}{8} + \frac{3a^4}{8} \cosh^{-1} \left(\frac{x}{a} \right).
\end{aligned}
$$

We shall now show what happens using the exponential definition of $\sinh t$. On the second line of the calculation we have used the binomial expansion.

$$
I = a^4 \int \sinh^4 t \, dt = a^4 \int \left(\frac{e^t - e^{-t}}{2} \right)^4 \, dt = \frac{a^4}{16} \int (e^t - e^{-t})^4 \, dt
$$

$$= \frac{a^4}{16} \int \left(e^{4t} - 4e^{2t} + 6 - 4e^{-2t} + e^{-4t} \right) dt$$

$$= \frac{a^4}{16} \left(\frac{e^{4t}}{4} - 2e^{2t} + 6t + 2e^{-2t} - \frac{e^{-4t}}{4} \right)$$

$$= \frac{a^4}{16} \left(\frac{\sinh 4t}{2} - 4\sinh 2t + 6t \right)$$

$$= a^4 \left(\frac{\sinh^3 t \cosh t}{4} - \frac{3 \sinh t \cosh t}{8} + \frac{3t}{8} \right).$$

This is the same expression as on the ninth line of the previous calculation, so the final steps will be identical.

It is occasionally worthwhile to undertake detailed calculations such as this by hand, but with a complicated integral there is a lot of potential for algebraic or arithmetic mistakes, and so MAPLE can be used to provide a reliable answer, or as a check on manual calculation. In this case the command

```
int((x^2-a^2)^(3/2),x);
```

gives exactly the answer we finished with. The command

```
int((sinh(t))^4,t);
```

gives the answer (without the a^4) on the line above containing $\sinh^3 t$ (the third line from the end of the first calculation).

Example 9.10

Evaluate the indefinite integral $I = \int x^2 \sqrt{a^2 - x^2} \, dx$ $(a^2 \geq x^2)$.

This integral seems to be similar to the previous example. However that involved $\sqrt{x^2 - a^2}$, and so required $x^2 \geq a^2$, whereas this example involves $\sqrt{a^2 - x^2}$, and so requires $a^2 \geq x^2$. The identity needed to get rid of the square root is the basic trigonometric one $\cos^2 t + \sin^2 t = 1$, corresponding to either of the substitutions $x = a \cos t$ or $x = a \sin t$. The corresponding hyperbolic identity would be $1 - \text{sech}^2 x = \tanh^2 x$. We shall use $x = a \sin t$; $dx = a \cos t \, dt$, showing the main steps and leaving the reader to verify the use of identities at each stage.

$$I = \int x^2 \sqrt{a^2 - x^2} \, dx = \int a^2 \sin^2 t \sqrt{a^2 (1 - \sin^2 t)} \, a \cos t \, dt$$

$$= a^4 \int \sin^2 t \cos^2 t \, dt = \frac{a^4}{4} \int \sin^2 2t \, dt = \frac{a^4}{8} \int (1 - \cos 4t) \, dt$$

$$= \frac{a^4}{8} \left(t - \frac{\sin 4t}{4} \right) = \frac{a^4}{8} \left(t - \sin t \cos t (1 - 2\sin^2 t) \right)$$

$$= \frac{a^4}{8}\left(\sin^{-1}\left(\frac{x}{a}\right) - \frac{x}{a}\sqrt{1 - \frac{x^2}{a^2}}\left(1 - 2\frac{x^2}{a^2}\right)\right)$$

$$= \frac{a^4}{8}\sin^{-1}\left(\frac{x}{a}\right) - \frac{1}{8}x\sqrt{a^2 - x^2}(a^2 - 2x^2).$$

The penultimate line is obtained by the method used in *STEP 7* of Example 9.7, involving a right-angled triangle.

9.4 Rational Functions of cos and sin

A rational function, as defined in Section 1.6.2, is a quotient of polynomials, where the numerator and denominator involve only terms containing non-negative integer powers of the variable x. By a rational function of cos and sin we mean a quotient where the numerator and denominator involve only terms containing non-negative integer powers of cos and sin, for example

$$\frac{\cos^3 t \sin t + \sin^2 t + 3 - 2\cos^2 t}{3\sin t - 4 + 5\sin^2 t \cos^5 t - 2\cos t}.$$

To integrate such a function we use the so-called **half-angle substitution** $x = \tan\left(\frac{t}{2}\right)$. This will always lead to a rational function of x. In Chapter 10 we shall consider methods of integrating rational functions in general, so the example in this section will be a straightforward one.

Because such expressions often involve both $\cos t$ and $\sin t$ we need to use identities which express these in terms of x. We can do this using a right-angled triangle together with basic trigonometric identities.

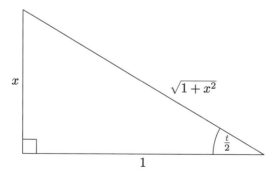

Figure 9.2   Half-angle substitution

In the right-angled triangle we have $x = \tan\left(\dfrac{t}{2}\right)$. So $dx = \dfrac{1}{2}\sec^2\left(\dfrac{t}{2}\right)dt$,

which we can rearrange in the form

$$dt = 2\cos^2\left(\frac{t}{2}\right)dx = \frac{2}{1+x^2}\,dx,$$

where we obtain the value of $\cos^2\left(\dfrac{t}{2}\right)$ from the triangle. We can also use the

triangle to see that

$$\sin\left(\frac{t}{2}\right) = \frac{x}{\sqrt{1+x^2}}.$$

We therefore have

$$\cos t \;=\; 2\cos^2\left(\frac{t}{2}\right) - 1 = \frac{2}{1+x^2} - 1 = \frac{1-x^2}{1+x^2},$$

$$\sin t \;=\; 2\sin\left(\frac{t}{2}\right)\cos\left(\frac{t}{2}\right) = 2\frac{x}{\sqrt{1+x^2}}\frac{1}{\sqrt{1+x^2}} = \frac{2x}{1+x^2}.$$

We can therefore see that each term in the integral of a rational function of $\cos t$ and $\sin t$ (including dt) will be transformed into a rational fraction of x, and so we shall be left with a rational function of x to integrate.

Example 9.11

Evaluate the indefinite integral $I = \displaystyle\int \frac{1}{2+\sin t}\,dt$.

Using the substitution $x = \tan\left(\dfrac{t}{2}\right)$ with the identities derived above gives

$$
\begin{aligned}
I \;&=\; \int \frac{1}{2+\sin t}\,dt = \int \frac{1}{2+\frac{2x}{1+x^2}}\frac{2}{1+x^2}\,dx \\[2mm]
&=\; \int \frac{1}{x^2+x+1}\,dx \quad \text{(after simplification)} \\[2mm]
&=\; \int \frac{1}{\left(x+\frac{1}{2}\right)^2 + \frac{3}{4}} \quad \text{(completing the square)} \\[2mm]
&=\; \frac{1}{\sqrt{\frac{3}{4}}}\tan^{-1}\left(\frac{x+\frac{1}{2}}{\sqrt{\frac{3}{4}}}\right) = \frac{2}{\sqrt{3}}\tan^{-1}\left(\frac{2x+1}{\sqrt{3}}\right) \\[2mm]
&=\; \frac{2}{\sqrt{3}}\tan^{-1}\left(\frac{2\tan\left(\frac{t}{2}\right)+1}{\sqrt{3}}\right).
\end{aligned}
$$

9.5 Substitution using MAPLE

We saw in Chapter 8 that MAPLE allows us to go through the procedure of integration by parts step-by-step, instead of simply providing the answer. Integration by substitution can be approached in a similar way.

Example 9.12

Use MAPLE to demonstrate the steps in Example 9.7.

We first need `with(student);` to load the appropriate package into memory. We now need to give the integral a name, using the command

```
M:=Int(sqrt(4*x^2-16*x+52),x);
```

(Note that we cannot use the symbol I for the integral in this case, as MAPLE reserves that for the complex number whose square is -1.)

We now complete the square, and MAPLE takes care of the fact that the coefficient of x^2 is not 1. So the command

```
completesquare(%,x);
```

gives the output

$$\int 2\sqrt{(x-2)^2 + 9}\,dx.$$

We now replace $(x - 2)$ by t, as in Example 9.7. MAPLE refers to substitution using the synonymous term "changing the variable". The command is

```
changevar(u=x-2,%,u);
```

where the final u tells MAPLE the variable of integration, since such a change of variable may contain other letters, for example $u = x - a$.

We have now reached the point where the integrand is the square root of a sum of two squares, and *STEP 5* of Example 9.7 tells us that we need to substitute $t = 3\tan u$, which is accomplished with the command

```
changevar(u=3*tan(t),%,t);
```

giving the output

$$\int \left(2\sqrt{9 + 9\tan(t)^2}(3 + 3\tan(t)^2)\right)dt.$$

We now need to simplify the integrand. Simplification of expressions in MAPLE has a variety of syntaxes depending on the context, and here we need

```
simplify(%,symbolic);
```

which produces

$$18\int \left((1 + \tan(t)^2)^{\frac{3}{2}}\right)dt.$$

We see that MAPLE does not use the required trigonometric identity, and a straightforward method of doing this is for us to realise ourselves that we have to make such a replacement, using

```
subs(1+tan(t)^2=sec(t)^2,%);
```

followed by another

```
simplify(%,symbolic);
```

to give

$$18 \int \frac{1}{\cos(t)^3} \, dt.$$

We have now reached *STEP 6* in Example 9.7, and the command

```
value(%)
```

gives

$$9\frac{\sin(t)}{\cos(t)^2} + 9\ln(\sec(t) + \tan(t)),$$

where again MAPLE has not done trigonometric simplification.

Finally we need to express the answer in terms of the original variable, so to substitute for t in terms of u we have to tell MAPLE to solve the equation $u = 3 \tan t$ and then replace t. We use the command

```
subs(t=solve(u=3*tan(t),t),%);
```

and obtain a complicated expression involving $\arctan(t)$. We need to apply a further

```
simplify(%,symbolic);
```

and then

```
subs(u=x-2,%);,
```

which gives the result

$$\sqrt{(x-2)^2 + 9} \, (x-2) - 9\ln 3 + 9\ln\left(\sqrt{(x-2)^2 + 9} + x - 2\right).$$

EXERCISES

9.1. Evaluate the following indefinite integrals, using an appropriate substitution.

(a) $\displaystyle\int 3x^2 \left(x^3 + 4\right)^{20} \, dx;$ (b) $\displaystyle\int x \left(x^2 - 6\right)^{\frac{4}{3}} \, dx;$

(c) $\displaystyle\int (x^2 + 1) \sqrt[3]{x^3 + 3x - 2} \, dx;$ (d) $\displaystyle\int \frac{x^2}{2x^3 + 5} \, dx;$

(e) $\displaystyle\int \frac{x^2}{4 + x^6} \, dx;$ (f) $\displaystyle\int \cos x \sin^7 x \, dx;$

(g) $\displaystyle\int \tan^5 x \sec^2 x \, dx;$ (h) $\displaystyle\int x^2 \cos(2x^3) \, dx;$

(i) $\displaystyle\int \frac{\sin\sqrt{x}}{\sqrt{x}}\,dx;$ (j) $\displaystyle\int xe^{x^2-2}\,dx;$

(k) $\displaystyle\int \frac{\cos(\ln x)}{x}\,dx;$ (l) $\displaystyle\int 2\sin x\cos x\,e^{\cos 2x}\,dx;$

(m) $\displaystyle\int \frac{e^x}{\sqrt{e^x+3}}\,dx;$ (n) $\displaystyle\int \frac{1}{e^x+2e^{-x}}\,dx.$

9.2. Evaluate the following definite integrals, using an appropriate substitution.

(a) $\displaystyle\int_0^1 x^2 e^{x^3}\,dx;$ (b) $\displaystyle\int_0^1 x^4(x^5-1)^6\,dx;$

(c) $\displaystyle\int_0^{\frac{\pi}{2}} \cos x\sqrt{\sin x}\,dx;$ (d) $\displaystyle\int_{\ln 7}^{\ln 26} e^x\sqrt[3]{1+e^x}\,dx.$

9.3. Evaluate the following indefinite integrals, using an appropriate substitution.

(a) $\displaystyle\int \frac{1}{\sqrt{-x^2-4x}}\,dx;$ (b) $\displaystyle\int \frac{1}{(3-x^2)^{\frac{3}{2}}}\,dx;$

(c) $\displaystyle\int \frac{x^2}{\sqrt{1-x^2}}\,dx;$ (d) $\displaystyle\int \frac{1}{x^2\sqrt{1-9x^2}}\,dx;$

(e) $\displaystyle\int \frac{1}{\sqrt{4+x^2}}\,dx;$ (f) $\displaystyle\int \frac{x}{\sqrt{2x-x^2}}\,dx;$

(g) $\displaystyle\int \frac{x^2}{(a^2-x^2)^{\frac{3}{2}}}\,dx;$ (h) $\displaystyle\int \sqrt{4x^2-8x+24}\,dx.$

9.4. Evaluate the following indefinite integrals, using the half-angle substitution.

(a) $\displaystyle\int \frac{1}{5+3\cos t}\,dt;$ (b) $\displaystyle\int \frac{1}{\cos t+\sin t+1}\,dt;$

(c) $\displaystyle\int_0^{\frac{\pi}{2}} \frac{1}{2+\cos t}\,dt;$ (d) $\displaystyle\int_0^{\frac{\pi}{2}} \frac{1}{3+3\cos t-\sin t}\,dt.$

9.5. Evaluate the following indefinite integrals.

(a) $\displaystyle\int \frac{\cos x}{\sqrt{1+4\sin x+\sin^2 x}}\,dx;$ (b) $\displaystyle\int \frac{e^x}{\sqrt{2-e^{2x}}}\,dx.$

10
Integration of Rational Functions

10.1 Introduction

A rational function is one of the form $R(x) = \dfrac{P(x)}{Q(x)}$, where are both $P(x)$ and $Q(x)$ are polynomials in the variable x, for example

$$\frac{2x^3 + 3x^2 - 4x + 1}{x^2 - 3x + 2}.$$

In this chapter we shall explain the steps involved in a procedure which will enable us to integrate any rational function, provided the algebra is not too horrible! One of the algebraic tools needed is the decomposition of rational functions into partial fractions, and we discuss this in the next section. In Section 10.3 we describe the process of integrating a rational function, split into a sequence of steps. We explain what happens at each step, using different examples at each stage to illustrate some degree of generality. In Section 10.4 we work through several examples showing the whole process applied to individual rational functions.

Apart from partial fractions, the main algebraic prerequisite is polynomial division, discussed in detail in Section 1.6.1.

10.2 Partial Fractions

The partial fraction decomposition expresses a rational function $\dfrac{P(x)}{Q(x)}$ as a sum of simpler algebraic fractions. The denominators of these fractions are determined by the factorisation of the denominator $Q(x)$. When a real polynomial is completely factorised into real factors, the factors will either be linear or quadratic, and some factors may occur more than once. Factorisation of polynomials was discussed in Section 1.6.1.

Dealing with partial fractions where the denominator factorises into linear factors only, none of which are repeated, is a familiar area of school mathematics. A common method used is that of equating coefficients, as in the following example.

Example 10.1

Find the partial fraction decomposition of

$$\frac{x^3 + 2x^2 - x + 4}{(x-1)(x+2)(x-3)(x+1)}.$$

The decomposition is of the form

$$\frac{x^3 + 2x^2 - x + 4}{(x-1)(x+2)(x-3)(x+1)} = \frac{A}{x-1} + \frac{B}{x+2} + \frac{C}{x-3} + \frac{D}{x+1}.$$

If we now put the right hand side over the common denominator

$$(x-1)(x+2)(x-3)(x+1),$$

the numerator will be

$$A(x+2)(x-3)(x+1) + B(x-1)(x-3)(x+1)$$
$$+C(x-1)(x+2)(x+1) + D(x-1)(x+2)(x-3).$$

So equating the numerators gives

$$x^3 + 2x^2 - x + 4 \equiv A(x+2)(x-3)(x+1) + B(x-1)(x-3)(x+1)$$
$$+C(x-1)(x+2)(x+1) + D(x-1)(x+2)(x-3),$$

where the use of the identity symbol $\equiv$ emphasises that this is true for all values of x. One method of finding A, B, C, D is to multiply out the right-hand side, giving

$$x^3 + 2x^2 - x + 4 \equiv (A + B + C + D)x^3 + (-3B + 2C - 2D)x^2$$
$$+(-7A - B - C - 5D)x + (-6A + 3B - 2C + 6D).$$

Because this is an identity the coefficient of each power of x on both sides must be the same. This gives the following system of four equations to solve for A, B, C, D.

$$
\begin{aligned}
A + B + C + D &= 1, \\
-3B + 2C - 2D &= 2, \\
-7A - B - C - 5D &= -1, \\
-6A + 3B - 2C + 6D &= 4.
\end{aligned}
$$

An alternative method is to substitute specific values of x in the identity, chosen so as to make one of the linear factors zero. This has the effect of making all but one of the terms zero on the right hand side of the first identity above. This enables us to determine the unknown coefficients one at a time. The results are given in the following table.

$x = 1$	$6 = -12A$	$A = -1/2$
$x = -2$	$6 = -15B$	$B = -2/5$
$x = 3$	$46 = 40C$	$C = 23/20$
$x = -1$	$6 = 8D$	$D = 3/4$

We therefore have

$$
\frac{x^3 + 2x^2 - x + 4}{(x-1)(x+2)(x-3)(x+1)} = -\frac{1}{2(x-1)} - \frac{2}{5(x+2)} + \frac{23}{20(x-3)} + \frac{3}{4(x+1)}.
$$

Example 10.2

In this example we show how MAPLE can be used to help with partial fraction decomposition. We have chosen an example where the degree of the numerator exceeds that of the denominator, and we can see that MAPLE performs polynomial division as well as partial fraction decomposition. The following are the relevant commands together with the outputs. Notice that the first command defines the function f. We can subsequently refer to f rather than having to retype it every time we wish to use it, as the second command shows.

```
f:=(x^4+x^3-2*x^2+3*x+3)/(x^2+x-2);
```

$$
f := \frac{x^4 + x^3 - 2x^2 + 3x + 3}{x^2 + x - 2}
$$

```
convert(f,parfrac,x);
```

$$
x^2 + \frac{1}{x+2} + \frac{2}{x-1}.
$$

In the command for converting to partial fractions we need the final x in case
there is another variable in the expression, for example if the coefficients were
letters instead of numbers.

When we have a rational function where the degree of the numerator is less
than the degree of the denominator, and decompose it into partial fractions,
the degree of the numerator will be less than the degree of the denominator in
each of the component fractions also. So if we have a partial denominator like
$(x-2)^4$ the degree of the corresponding numerator in the decomposition will
be less than 4, so it will be cubic, quadratic, linear, or a constant. If we have a
partial denominator like $(x^2 + x + 1)^3$, where the quadratic has no real factors,
in this case its degree is 6, so the associated numerator has degree at most 5.

In the case of fractions with repeated factors in the denominator, we can
decompose them further. In the case of a linear factor like $(ax + b)$ the decomposition is

$$\frac{p(x)}{(ax+b)^n} = \frac{c_1}{(ax+b)} + \frac{c_2}{(ax+b)^2} + \frac{c_3}{(ax+b)^3} + \cdots + \frac{c_n}{(ax+b)^n},$$

where $p(x)$ is the numerator (of degree less than n) and the c_i are all constants.
This is illustrated in Example 10.3.

In the case of a repeated quadratic factor the decomposition is

$$\frac{s(x)}{(ax^2+bx+c)^n} = \frac{c_1 x + d_1}{(ax^2+bx+c)} + \frac{c_2 x + d_2}{(ax^2+bx+c)^2} + \cdots + \frac{c_n x + d_n}{(ax^2+bx+c)^n},$$

where $s(x)$ is the numerator (of degree less than $2n$) and the c_i and d_i are
all constants. This is illustrated in Example 10.6. These results are sometimes
proved in algebra textbooks. We outline a method for proving the result for
linear denominators in the exercises at the end of this chapter using Taylor's
Theorem, encountered in Chapter 6.

Example 10.3

Find the partial fraction decomposition of

$$\frac{x^2 + x}{(x-1)^3}.$$

The decomposition will be of the form

$$\frac{x^2 + x}{(x-1)^3} = \frac{A}{x-1} + \frac{B}{(x-1)^2} + \frac{C}{(x-1)^3}.$$

In this example we use a method of compensation whereby we replace x
throughout by the factor in the denominator, $(x - 1)$ in this case, and then

compensate the constants. This is an alternative to the method of comparing coefficients described in Example 10.1. It is a device used in various algebraic calculations.

$$\frac{x^2 + x}{(x-1)^3} = \frac{[(x-1)+1]^2 + [(x-1)+1]}{(x-1)^3},$$

$$\frac{(x-1)^2 + 3(x-1) + 2}{(x-1)^3} = \frac{1}{(x-1)} + \frac{3}{(x-1)^2} + \frac{2}{(x-1)^3}.$$

Example 10.4

Find the partial fraction decomposition of

$$\frac{x^3}{(x^2 + x + 1)^2}.$$

In this example we have a repeated quadratic factor in the denominator. In such cases the numerators are always linear (or constant if the coefficient of x turns out to be zero). We use a method of compensation similar to the previous example. We regard the numerator x^3 as $x.x^2$, and we replace the second factor x^2 by the quadratic $x^2 + x + 1$ which occurs in the denominator, and then compensate.

$$\frac{x^3}{(x^2 + x + 1)^2} = \frac{x(x^2 + x + 1) - (x^2 + x + 1) + 1}{(x^2 + x + 1)^2}$$

$$= \frac{x-1}{(x^2 + x + 1)} + \frac{1}{(x^2 + x + 1)^2}.$$

Example 10.5

Decompose $\dfrac{x^2 - 2x + 3}{(x+1)^3}$ into partial fractions.

The decomposition will be of the form

$$\frac{x^2 - 2x + 3}{(x+1)^3} = \frac{A}{x+1} + \frac{B}{(x+1)^2} + \frac{C}{(x+1)^3}.$$

We have to find A, B, C and as usual we arrange the right-hand side as a single fraction over the common denominator, $(x+1)^3$. This gives rise to the numerator $A(x+1)^2 + B(x+1) + C$, which must be equal to $x^2 - 2x + 3$. We could find A, B, C, by comparing coefficients, as in Example 10.1, or by compensation, as in Example 10.3. Here we shall demonstrate another method, involving differentiation, reminiscent of the discussion in Section 6.4 on Taylor Polynomials.

So in the equation $A(x+1)^2 + B(x+1) + C = x^2 - 2x + 3$, we substitute $x = -1$ to give $C = 6$. We then differentiate both sides of the equation to give $2A(x+1) + B = 2x - 2$. Putting $x = -1$ now gives $B = -4$. Differentiating once more gives $2A = 2$, and so $A = 1$. Therefore the decomposition is

$$\frac{x^2 - 2x + 3}{(x+1)^3} = \frac{1}{x+1} - \frac{4}{(x+1)^2} + \frac{6}{(x+1)^3}.$$

Example 10.6

Decompose $\dfrac{2x^2 - x + 4}{(x^2 + x + 1)^2}$ into partial fractions.

The decomposition will be of the form $\dfrac{Ax + B}{(x^2 + x + 1)} + \dfrac{Cx + D}{(x^2 + x + 1)^2}.$

Arranging this as a single fraction over a common denominator gives the numerator $(Ax + B)(x^2 + x + 1) + Cx + D$, which must be equal to $2x^2 - x + 4$.

We can't use the differentiation method in the same way as in Example 10.5, because there is a quadratic involved, and indeed there is no real value of x for which it is zero. We can adapt the differentiation method however, using $x = 0$ in each case to simplify the calculations. So we have to solve the identity $(Ax+B)(x^2+x+1)+Cx+D \equiv 2x^2-x+4$ for A, B, C. We can see immediately that $A = 0$ because there is no x^3 term on the right hand side. So the identity simplifies to $B(x^2 + x + 1) + Cx + D \equiv 2x^2 - x + 4$. Putting $x = 0$ gives $B + D = 4$.

Differentiating gives $2Bx + B + C \equiv 4x - 1$, and putting $x = 0$ gives $B + C = -1$.

Differentiating again gives $2B = 4$ and so $B = 2$. We can now deduce that $C = -3$ and $D = 2$. So the decomposition is

$$\frac{2x^2 - x + 4}{(x^2 + x + 1)^2} = \frac{2}{(x^2 + x + 1)} + \frac{-3x + 2}{(x^2 + x + 1)^2}.$$

Notice that although we expect each numerator to be linear, it is possible that some of the coefficients may be zero, as has happened with the first fraction in this case.

In the examples above we have used a variety of methods, chosen to keep the algebraic manipulation as straightforward as we can.

10.3 The Integration Process

Integrating a rational function $\dfrac{P(x)}{Q(x)}$ is a process which can be broken down into a well-defined sequence of steps. In this section we shall describe the procedure, providing illustrative examples. In the subsequent section we work through some examples in detail.

STEP 1 Polynomial Division

Is the degree of $P(x)$ greater than or equal to the degree of $Q(x)$? If the answer is YES then divide $Q(x)$ into $P(x)$ to obtain $\dfrac{P(x)}{Q(x)} = A(x) + \dfrac{B(x)}{Q(x)}$, where $A(x)$ and $B(x)$ are polynomials and $\deg B(x) < \deg Q(x)$. Polynomial division was discussed in detail in Section 1.6.1.

STEP 2 Factorisation

Following polynomial division we need to factorise the denominator $Q(x)$. The problem with this step is that there is no general algorithm which will factorise all polynomials, as explained in Section 1.6.1. So in practice this step can only be carried out if the polynomial $Q(x)$ is relatively straightforward.

STEP 3 Partial Fractions

Decompose $\dfrac{B(x)}{Q(x)}$ using partial fractions. This was discussed in detail in Section 10.2.

STEP 4 Integration

We can now integrate each term in the partial fraction decomposition separately. Each is a rational function, but only a few different types of expression occur, as we have seen in Section 10.2, and we discuss each of them. The first two types involve a linear factor in the denominator, which may be repeated, and a constant numerator. From the basic integrals described in Section 7.2 we have the following two results.

$$\int \frac{A}{x+k}\,dx = A\ln|x+k|,$$

$$\int \frac{A}{(x+k)^n}\,dx = A\frac{(x+k)^{-n+1}}{-n+1} \quad (n \neq 1).$$

We now need to consider quadratic denominators. Where the quadratic is not a repeated factor the integral will be of the form $\displaystyle\int \frac{\text{linear}}{\text{quadratic}}$.

We can re-write the integrand in the form $\dfrac{\text{linear}}{q(x)} = c\dfrac{q'(x)}{q(x)} + \dfrac{d}{q(x)}$, where c and d are constants, and integrate each term separately.

Example 10.7

Evaluate the integral $\displaystyle\int \frac{x+3}{x^2-x+4}\,dx$.

We rewrite the integrand as explained above to give

$$\int \frac{x+3}{x^2-x+4}\,dx = \int \frac{\frac{1}{2}(2x-1)}{x^2-x+4}\,dx + \int \frac{\frac{7}{2}}{x^2-x+4}\,dx.$$

Using the general result

$$\int \frac{q'(x)}{q(x)}\,dx = \ln|q(x)|,$$

we can deal with the first integral as follows.

$$\int \frac{\frac{1}{2}(2x-1)}{x^2-x+4}\,dx = \frac{1}{2}\ln|x^2-x+4|.$$

We deal with the second integral by completing the square of the denominator, which doesn't have real roots.

$$\int \frac{\frac{7}{2}}{x^2-x+4}\,dx = \frac{7}{2}\int \frac{dx}{\left(x-\frac{1}{2}\right)^2+\frac{15}{4}} = \frac{7}{2}\sqrt{\frac{4}{15}}\,\tan^{-1}\left(\sqrt{\frac{4}{15}}\left(x-\frac{1}{2}\right)\right).$$

Finally we must deal with integrals of the form $\displaystyle\int \frac{\text{linear}}{(\text{quadratic})^n}$ $\quad(n>1)$.

We can write the integrand as $\dfrac{\text{linear}}{(q(x))^n} = c\dfrac{q'(x)}{(q(x))^n} + \dfrac{d}{(q(x))^n}$, where c and d are constants.

The following example illustrates the general process for evaluating such integrals.

Example 10.8

Show how to evaluate the integral

$$\int \frac{2x-1}{(x^2-2x+5)^n}\,dx \quad (n>1).$$

We first split the integral into two parts, as explained above. This gives

$$\int \frac{2x-1}{(x^2-2x+5)^n}\,dx = \int \frac{2x-2}{(x^2-2x+5)^n}\,dx + \int \frac{1}{(x^2-2x+5)^n}\,dx.$$

We can deal with the first integral as follows

$$\int \frac{2x-2}{(x^2-2x+5)^n}\,dx = \frac{\left(x^2-2x+5\right)^{-n+1}}{-n+1}.$$

We deal with the second integral by completing the square of the denominator, which doesn't have real roots, and then let $u = x - 1$.

$$\int \frac{1}{(x^2 - 2x + 5)^n} \, dx = \int \frac{1}{((x-1)^2 + 4)^n} \, dx = \int \frac{1}{(u^2 + 4)^n} \, du.$$

We now use either the trigonometric substitution

$$u = 2 \tan \theta; \, du = 2 \sec^2 \theta \, d\theta,$$

or the hyperbolic substitution

$$u = 2 \sinh t; \, du = 2 \cosh t \, dt.$$

Using the trigonometric substitution gives

$$\int \frac{1}{(u^2 + 4)^n} \, du = \int \frac{2 \sec^2 \theta}{(4 \sec^2 \theta)^n} \, d\theta = \frac{2}{4^n} \int (\cos \theta)^{2n-2} \, d\theta,$$

for which a reduction formula is needed (see Section 8.2).

Using the hyperbolic substitution gives

$$\int \frac{1}{(u^2 + 4)^n} \, du = \int \frac{2 \cosh t}{(4 \cosh^2 t)^n} \, dt = \frac{2}{4^n} \int (\cosh t)^{1-2n} \, dt,$$

and again a reduction formula is needed.

This procedure is illustrated in the next example, taking $n = 2$. In this case we do not need a reduction formula since n is sufficiently small to work out the relevant integrals directly.

Example 10.9

Evaluate the integral

$$\int \frac{2x - 1}{(x^2 - 2x + 5)^2} \, dx.$$

Splitting the integral as explained above gives

$$\int \frac{2x - 1}{(x^2 - 2x + 5)^2} \, dx = \int \frac{2x - 2}{(x^2 - 2x + 5)^2} \, dx + \int \frac{1}{(x^2 - 2x + 5)^2} \, dx.$$

Evaluating the first integral gives

$$\int \frac{2x - 2}{(x^2 - 2x + 5)^2} \, dx = -\frac{1}{x^2 - 2x + 5}.$$

The second integral can be evaluated as follows, using the trigonometric substitution $u = 2\tan t$, $du = 2\sec^2 t\, dt$.

$$
\begin{aligned}
\int \frac{1}{(x^2 - 2x + 5)^2}\, dx &= \int \frac{1}{((x-1)^2 + 4)^2}\, dx = \int \frac{1}{(u^2 + 4)^2}\, du \\
&= \int \frac{2\sec^2 t}{(4\tan^2 t + 4)^2}\, dt = \int \frac{2\sec^2 t}{(4\sec^2 t)^2}\, dt \\
&= \frac{1}{8} \int \cos^2 t\, dt = \frac{1}{16} \int (1 + \cos 2t)\, dt \\
&= \frac{1}{16} \left(t + \frac{\sin 2t}{2} \right) = \frac{1}{16}(t + \sin t \cos t) \\
&= \frac{1}{16} \left(\tan^{-1}\left(\frac{u}{2}\right) + \frac{u}{\sqrt{u^2 + 4}} \frac{2}{\sqrt{u^2 + 4}} \right) \\
&= \frac{1}{16} \left(\tan^{-1}\left(\frac{x-1}{2}\right) + \frac{2(x-1)}{x^2 - 2x + 5} \right),
\end{aligned}
$$

where in the penultimate line we have used the right-angled triangle method explained in Example 9.7.

Adding the two results together then gives

$$
\int \frac{2x - 1}{(x^2 - 2x + 5)^2}\, dx = \frac{1}{16} \left(\tan^{-1}\left(\frac{x-1}{2}\right) + \frac{2(x-9)}{x^2 - 2x + 5} \right).
$$

10.4 Examples

In this section we shall work through some examples, to help consolidate the general explanation in Section 10.3.

Example 10.10

Evaluate the indefinite integral

$$
\int \frac{10x^2 + 2x - 9}{2x^3 + x^2 - 9}\, dx.
$$

The degree of the numerator is less than that of the denominator, so we proceed directly to *STEP 2*, factorising the denominator. Trying the integer factors of the constant term, $(1, -1, 3, -3, 9, -9)$ does not work. However we note that the coefficient of the highest power of x is 2, and so we should try each of the integer factors divided by 2. Substituting these in turn tells us that $x = \frac{3}{2}$ is a root of the cubic, and so $(2x - 3)$ will be a factor. We can therefore find the remaining factor by dividing the cubic by $2x - 3$, using polynomial division, or alternatively by writing

$$2x^3 + x^2 - 9 = (2x - 3)(Px^2 + Qx + R).$$

Comparing coefficients tells us that on the right-hand side P must be 1 and R must be 3. We therefore have

$$2x^3 + x^2 - 9 = (2x - 3)(x^2 + Qx + 3).$$

The coefficient of x^2 on the right-hand side is $2Q - 3$, which must be equal to 1, giving $Q = 2$. We therefore deduce that

$$\frac{10x^2 + 2x - 9}{2x^3 + x^2 - 9} = \frac{10x^2 + 2x - 9}{(2x - 3)(x^2 + 2x + 3)}.$$

The quadratic factor in the denominator has no real roots, so we proceed to *STEP 3*, finding the partial fraction decomposition, which will be of the form

$$\frac{10x^2 + 2x - 9}{2x^3 + x^2 - 9} = \frac{A}{2x - 3} + \frac{Bx + C}{x^2 + 2x + 3}.$$

Recombining the right-hand side over the common denominator and then equating the numerators gives

$$10x^2 + 2x - 9 = A(x^2 + 2x + 3) + (Bx + C)(2x - 3).$$

Of the various methods available we will compare coefficients. This gives the three simultaneous equations

$$
\begin{aligned}
A + 2B &= 10, \\
2A + 2C - 3B &= 2, \\
3A - 3C &= -9.
\end{aligned}
$$

Solving these three equations gives $A = 2, B = 4, C = 5$. Therefore

$$\frac{10x^2 + 2x - 9}{2x^3 + x^2 - 9} = \frac{2}{2x - 3} + \frac{4x + 5}{x^2 + 2x + 3}.$$

We can now implement *STEP 4* and integrate. Firstly we must rearrange the second term in a suitable form, as in Example 10.7. We then have

$$
\begin{aligned}
\int \frac{10x^2 + 2x - 9}{2x^3 + x^2 - 9}\, dx &= \int \frac{2}{2x - 3} + \frac{4x + 5}{x^2 + 2x + 3}\, dx \\
&= \int \frac{2}{2x - 3}\, dx + 2 \int \frac{2x + 2}{x^2 + 2x + 3}\, dx + \int \frac{1}{x^2 + 2x + 3}\, dx \\
&= \int \frac{2}{2x - 3}\, dx + 2 \int \frac{2x + 2}{x^2 + 2x + 3}\, dx + \int \frac{1}{(x + 1)^2 + 2}\, dx \\
&= \ln|2x - 3| + 2\ln|x^2 + 2x + 3| + \frac{1}{\sqrt{2}} \tan^{-1}\left(\frac{x + 1}{\sqrt{2}}\right).
\end{aligned}
$$

Example 10.11

Evaluate the indefinite integral

$$\int \frac{8x^4 + 12x^3 + 7x^2 + 2x - 2}{8x^3 + 12x^2 + 6x + 1} \, dx.$$

The degree of the numerator exceeds that of the denominator, so *STEP 1* must be implemented. Following Example 1.12, for the first stage we write

$$8x^4 + 12x^3 + 7x^2 + 2x - 2 = x(8x^3 + 12x^2 + 6x + 1) + a(x).$$

This simplifies to give

$$7x^2 + 2x - 2 = 6x^2 + x + a(x),$$

and so $a(x) = x^2 + x - 2$. Therefore

$$\frac{8x^4 + 12x^3 + 7x^2 + 2x - 2}{8x^3 + 12x^2 + 6x + 1} = x + \frac{x^2 + x - 2}{8x^3 + 12x^2 + 6x + 1}.$$

For *STEP 2* we have to factorise the denominator, and acquaintance with binomial expansions should enable us to recognise that

$$8x^3 + 12x^2 + 6x + 1 = (2x + 1)^3.$$

The partial fraction decomposition needed for *STEP 3* has the form

$$\frac{x^2 + x - 2}{(2x + 1)^3} = \frac{A}{(2x + 1)} + \frac{B}{(2x + 1)^2} + \frac{C}{(2x + 1)^3}.$$

Putting the right-hand side over the common denominator and equating numerators gives

$$x^2 + x - 2 = A(2x + 1)^2 + B(2x + 1) + C.$$

We shall use the method involving differentiation introduced in Example 10.5. Firstly putting $x = -\frac{1}{2}$ will give $C = -\frac{9}{4}$.

Differentiating gives $2x + 1 = 2B + 4A(2x + 1)$, and substituting $x = -\frac{1}{2}$ shows that $B = 0$. Finally, differentiating again gives $A = \frac{1}{4}$. Therefore

$$\frac{x^2 + x - 2}{(2x + 1)^3} = \frac{1}{4(2x + 1)} - \frac{9}{4(2x + 1)^3}.$$

We are now in a position to integrate (*STEP 4*) without any further algebra.

$$\int \frac{8x^4 + 12x^3 + 7x^2 + 2x - 2}{8x^3 + 12x^2 + 6x + 1} \, dx$$

$$= \int x \, dx + \int \frac{1}{4(2x + 1)} \, dx - \int \frac{9}{4(2x + 1)^3} \, dx$$

$$= \frac{x^2}{2} + \frac{\ln|2x + 1|}{8} + \frac{9}{16(2x + 1)^2}.$$

Example 10.12

Evaluate the indefinite integral

$$I = \frac{4x^3 + 34x^2 + 120x + 127}{(2x^2 + 12x + 26)^2} \, dx.$$

Here the degree of the numerator is less than that of the denominator. Furthermore the quadratic in the denominator has no real roots. The first calculation involved is therefore *STEP 3*, the partial fraction decomposition. Since we have a repeated quadratic denominator, this decomposition has the form

$$\frac{4x^3 + 34x^2 + 120x + 127}{(2x^2 + 12x + 26)^2} = \frac{Ax + B}{2x^2 + 12x + 26} + \frac{Cx + D}{(2x^2 + 12x + 26)^2}.$$

Putting the right-hand side over the common denominator and equating numerators gives, after multiplying out,

$$2Ax^3 + (12A + 2B)x^2 + (26A + 12B + C)x + (26B + D)$$
$$\equiv 4x^3 + 34x^2 + 120x + 127.$$

Comparing coefficients works well in this case, and we see that

from the coefficient of x^3:　　$A = 2$,

from the coefficient of x^2:　　$24 + 2B = 34$,　giving　$B = 5$,

from the coefficient of x:　　$52 + 60 + C = 120$,　giving　$C = 8$,

from the constant term:　　$130 + D = 127$,　giving　$D = -3$.

Therefore the partial fraction decomposition is

$$\frac{4x^3 + 34x^2 + 120x + 127}{(2x^2 + 12x + 26)^2} = \frac{2x + 5}{2x^2 + 12x + 26} + \frac{8x - 3}{(2x^2 + 12x + 26)^2}.$$

We shall integrate each term separately (*STEP 4*). Firstly

$$\int \frac{2x + 5}{2x^2 + 12x + 26} \, dx = \frac{1}{2} \int \frac{2x + 6}{x^2 + 6x + 13} \, dx - \frac{1}{2} \int \frac{1}{x^2 + 6x + 13} \, dx.$$

Completing the square in the second integral gives

$$\int \frac{2x + 5}{2x^2 + 12x + 26} \, dx = \frac{1}{2} \int \frac{2x + 6}{x^2 + 6x + 13} \, dx - \frac{1}{2} \int \frac{1}{(x + 3)^2 + 4} \, dx.$$

Each integral is now in a standard form which enables us to write down the answer, as in Example 10.7.

$$\int \frac{2x + 5}{2x^2 + 12x + 26} \, dx = \frac{1}{2} \ln |x^2 + 6x + 13| - \frac{1}{4} \tan^{-1} \left(\frac{x + 3}{2} \right).$$

We now have to evaluate

$$\int \frac{8x - 3}{(2x^2 + 12x + 26)^2} \, dx.$$

We rearrange the integrand following the procedure of Example 10.9.

$$\frac{8x - 3}{(2x^2 + 12x + 26)^2} = \frac{2x + 6}{(x^2 + 6x + 13)^2} - \frac{27}{4((x + 3)^2 + 4)^2}.$$

We integrate each term separately, the first being straightforward, as in Example 10.9.

$$\int \frac{2x + 6}{(x^2 + 6x + 13)^2} \, dx = -\frac{1}{x^2 + 6x + 13}.$$

For the second term we need to use the substitution $x + 3 = 2\tan t$, which gives

$$\int \frac{27}{4((x + 3)^2 + 4)^2} \, dx = \frac{27}{4} \int \frac{2\sec^2 t}{16(\sec^2 t)^2} \, dt = \frac{27}{32} \int \cos^2 t \, dt$$

$$= \frac{27}{64}(t + \sin t \cos t) = \frac{27}{64}\left(\tan^{-1}\left(\frac{x + 3}{2}\right) + \frac{2(x + 3)}{x^2 + 6x + 13}\right),$$

where we have used the right-angled triangle method introduced in Example 9.7 to express $\cos t$ and $\sin t$ in terms of x. Finally we assemble all the results to give

$$\begin{aligned}
I &= \int \frac{4x^3 + 34x^2 + 120x + 127}{(2x^2 + 12x + 26)^2} \, dx \\
&= \frac{1}{2}\ln|x^2 + 6x + 13| - \frac{1}{4}\tan^{-1}\left(\frac{x + 3}{2}\right) - \frac{1}{x^2 + 6x + 13} \\
&\quad -\frac{27}{64}\left(\frac{2(x + 3)}{x^2 + 6x + 13} - \tan^{-1}\left(\frac{x + 3}{2}\right)\right) \\
&= \frac{1}{2}\ln|x^2 + 6x + 13| - \frac{43}{64}\tan^{-1}\left(\frac{x + 3}{2}\right) - \frac{27x + 113}{32(x^2 + 6x + 13)},
\end{aligned}$$

which can be checked using MAPLE.

Example 10.13

Find the indefinite integral of

$$f(x) = \frac{6x^6 + 37x^5 + 145x^4 + 338x^3 + 374x^2 + 70x + 38}{3x^5 + 17x^4 + 60x^3 + 122x^2 + 72x - 40}.$$

This is algebraically more complicated than the previous example, using polynomial division and a more involved partial fraction decomposition. The algebraic manipulation can be done using MAPLE, and so has not been given in detail. You would not expect to do as complicated an example as this by hand.

Make sure you understand all the algebraic procedures in STEPS 1–3, and that you understand each separate integral in STEP 4.

STEP 1

Divide the denominator into the numerator to give

$$f(x) = 1 + 2x + \frac{2(4x^4 + 17x^3 + 54x^2 + 39x + 39)}{3x^5 + 17x^4 + 60x^3 + 122x^2 + 72x - 40}.$$

STEP 2

Factorise the denominator

$$f(x) = 1 + 2x + \frac{2(4x^4 + 17x^3 + 54x^2 + 39x + 39)}{(3x - 1)(x + 2)^2(x^2 + 2x + 10)}.$$

STEP 3

Decompose into partial fractions

$$f(x) = 1 + 2x + \frac{2}{3x - 1} + \frac{x - 1}{(x + 2)^2} + \frac{x + 3}{x^2 + 2x + 10}.$$

As we saw in Example 10.2, MAPLE will perform the first three steps together. If we let $F(x)$ denote the integrand, then this can be accomplished using the command

```
convert(F(x),parfrac,x);
```

STEP 4

To integrate each term we need to rearrange the fractions in a suitable form for integration using the guidance in the examples above.

$$f(x) = 1 + 2x + \frac{2}{3}\frac{3}{3x - 1} + \frac{(x + 2) - 3}{(x + 2)^2} + \frac{\frac{1}{2}(2x + 2) + 2}{x^2 + 2x + 10}$$

$$= 1 + 2x + \frac{2}{3}\frac{3}{3x - 1} + \frac{1}{x + 2} - \frac{3}{(x + 2)^2} + \frac{1}{2}\frac{(2x + 2)}{x^2 + 2x + 10} + \frac{2}{(x + 1)^2 + 9}.$$

We can now integrate each term to obtain

$$\int f(x)\,dx = x + x^2 + \frac{2}{3}\ln|3x - 1| + \ln|x + 2| + \frac{3}{x + 2}$$

$$+ \frac{1}{2}\ln|x^2 + 2x + 10| + \frac{2}{3}\tan^{-1}\frac{x + 1}{3}.$$

EXERCISES

10.1. Evaluate the indefinite integrals of the rational functions defined by the following expressions.

(a) $\dfrac{7}{x^2 - 5x - 6}$;

(b) $\dfrac{3x - 11}{x^2 - 5x + 6}$;

(c) $\dfrac{3}{x^2 + 4x + 13}$;

(d) $\dfrac{2x + 3}{x^2 + 6x + 10}$;

(e) $\dfrac{x^2 + 2x - 3}{x^2 - 4x + 5}$;

(f) $\dfrac{x^3 + 2x^2 - x - 1}{x^2 + 6x + 13}$;

(g) $\dfrac{3x - 1}{x^2 + 4x + 4}$;

(h) $\dfrac{5x^2 + 4}{x^3 - 3x^2 + 3x - 1}$;

(i) $\dfrac{2x^2 - 11}{x^2 + 6x + 9}$;

(j) $\dfrac{x^2 + 1}{x^3 + x^2 + 3x - 5}$;

(k) $\dfrac{x^3 - x^2 + 4}{x^3 + x^2 + 3x - 5}$;

(l) $\dfrac{25x}{x^4 - x^2 - 2x + 2}$;

(m) $\dfrac{2x^2 - 3x + 4}{x^4 - 2x^2 + 1}$;

(n) $\dfrac{x^2 + 3x - 2}{(x^2 - 4x + 5)^2}$.

10.2. Evaluate the indefinite integral of the rational function defined by the following expression.

$$\frac{2x^4 - 14x^3 + 51x^2 - 89x + 82}{(x - 1)(x^2 - 4x + 7)^2}.$$

Use MAPLE to perform the partial fraction decomposition, and then integrate each term separately, as in Example 10.13.

10.3. This exercise provides a proof of the partial fraction decomposition for a repeated linear denominator. So we write

$$\frac{P(x)}{(ax + b)^n} = \frac{c_1}{(ax + b)} + \frac{c_2}{(ax + b)^2} + \frac{c_3}{(ax + b)^3} + \cdots + \frac{c_n}{(ax + b)^n},$$

where $P(x)$ denotes a polynomial of degree less than n.

Apply Taylor's Theorem to $P(x)$ about $x = -b/a$, and explain why the error term $E_k(x)$ is zero for $k \geq n - 1$. Hence show that $P(x)$ can be expressed as $Q\left(x + \dfrac{b}{a}\right)$, where Q is a polynomial of degree at most $n - 1$.

Deduce that $P(x)$ can be expressed as $R(ax + b)$, where R is a polynomial of degree at most $n - 1$.

Divide both sides of the equation $P(x) = R(ax + b)$ by $(ax + b)^n$ to obtain the partial fraction decomposition.

11

Geometrical Applications of Integration

In Section 7.1 we discussed integration as summation, and we use that interpretation of the integral in this chapter to construct integral formulae for some geometrical quantities. We shall consider length, area and volume, and the notions of centroid and centre of mass. The most important thing in this chapter is not to remember particular formulae, but to understand the principles underpinning their construction, so that analogous formulae can be constructed in other areas of application.

11.1 Arc Length

In this section we derive formulae for the length of a curve. Not all curves can be given as the graph of a function, for example a circle, and so we shall deal with the more general situation where a curve is described parametrically by means of the equations

$$x = x(t), \quad y = y(t), \quad a \leq t \leq b.$$

We shall assume that we have a smooth curve, for which the functions $x(t)$ and $y(t)$ have continuous derivatives for $a \leq t \leq b$. On the graph represented by these parametric equations, we divide the curve into small pieces by means of a sequence of points $P_0, P_1, P_2, \ldots, P_n$, specified by the sequence of values of the parameter, given by

$$a = t_0 < t_1 < t_2 < \ldots < t_n = b.$$

This is shown in the left-hand diagram in Figure 11.1. In the right-hand diagram we have isolated one small piece of the graph lying between two successive points. The length of that small piece of arc is denoted conventionally by ds.

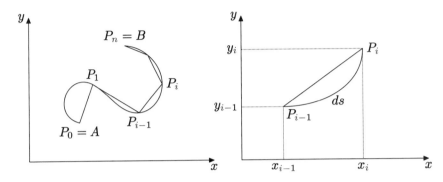

Figure 11.1   Arc length

In the right-hand diagram, if the piece of arc is very small then the gradient will not change much along the arc because the parametric functions are assumed to have continuous derivatives. So the length ds will be approximately equal to that of the line segment $P_{i-1}P_i$. Using Pythagoras' Theorem gives

$$
\begin{aligned}
(P_{i-1}P_i)^2 &= (x_i - x_{i-1})^2 + (y_i - y_{i-1})^2 \\
&= (x(t_i) - x(t_{i-1}))^2 + (y(t_i) - y(t_{i-1}))^2 \\
&= (x'(c_i)(t_i - t_{i-1}))^2 + (y'(d_i)(t_i - t_{i-1}))^2,
\end{aligned}
$$

using the Mean Value Theorem (Theorem 6.1).

We now deduce that

$$
ds^2 \approx (P_{i-1}P_i)^2 = \left(x'(c_i)^2 + y'(d_i)^2\right)(t_i - t_{i-1})^2.
$$

Taking square roots gives

$$
ds \approx \sqrt{(x'(c_i)^2 + y'(d_i)^2)}\,(t_i - t_{i-1}).
$$

The total arc length is therefore given by

$$
L \approx \sum_{i=1}^{n} \sqrt{(x'(c_i)^2 + y'(d_i)^2)}\,(t_i - t_{i-1}).
$$

This is an approximating sum to an integral, as outlined in Section 7.1, and so we finally we have the formula

$$
L = \int_a^b \sqrt{(x'(t)^2 + y'(t)^2)}\,dt.
$$

As a special case, suppose that the curve is the graph of the function specified by $y = f(x)$, $p \leq x \leq q$. This can be expressed in the parametric form $x = t$, $y = f(t)$, $p \leq t \leq q$. We then have

$$x'(t) = 1, \quad y'(t) = f'(t) = f'(x) = \frac{dy}{dx}.$$

So the formula becomes

$$L = \int_p^q \sqrt{(1 + f'(x)^2)}\, dx = \int_p^q \sqrt{1 + \left(\frac{dy}{dx}\right)^2}\, dx.$$

Example 11.1

Find the length of the curve given by $x = t^2$, $y = 2t^3$, $-1 \leq t \leq 1$, shown in Figure 11.2.

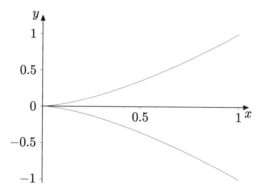

Figure 11.2   Graph of $x = t^2$, $y = 2t^3$, $-1 \leq t \leq 1$

Using the formula derived above gives

$$L = \int_{-1}^1 \sqrt{4t^2 + 36t^4}\, dt = \int_{-1}^1 2t\sqrt{1 + 9t^2}\, dt = \left[\frac{2}{3}\frac{1}{9}\left(1 + 9t^2\right)^{\frac{3}{2}}\right]_{-1}^1 = 0.$$

This is clearly wrong. The curve does not have zero length! The problem is that we were not sufficiently careful with the square root when we factored out t^2 from under the square root in the first integral. It is not true in general that $\sqrt{t^2} = t$, especially in this case where the integrand must be positive for all values of t because it represents a length.

A correct version of the calculations is as follows.

$$L = \int_{-1}^1 \sqrt{4t^2 + 36t^4}\, dt = \int_{-1}^1 2|t|\sqrt{1 + 9t^2}\, dt$$

$$= 2 \int_0^1 2t\sqrt{1 + 9t^2}\, dt = 2 \left[\frac{2}{3}\frac{1}{9}\left(1 + 9t^2\right)^{\frac{3}{2}} \right]_0^1$$

$$= \frac{4}{27}\left(10^{\frac{3}{2}} - 1\right).$$

Example 11.2

Find the length of the curve given by $y = x^2 - \dfrac{\ln x}{8}$ $(1 \le x \le 2)$.

In this example we need to use the cartesian formula for arc length.

$$L = \int_1^2 \sqrt{1 + \left(\frac{dy}{dx}\right)^2}\, dx = \int_1^2 \sqrt{1 + \left(2x - \frac{1}{8x}\right)^2}\, dx$$

$$= \int_1^2 \sqrt{1 + 4x^2 + \frac{1}{64x^2} - \frac{1}{2}}\, dx = \int_1^2 \sqrt{4x^2 + \frac{1}{64x^2} + \frac{1}{2}}\, dx$$

$$= \int_1^2 \sqrt{\left(2x + \frac{1}{8x}\right)^2}\, dx = \int_1^2 \left(2x + \frac{1}{8x}\right) dx$$

$$= \left[x^2 + \frac{\ln x}{8}\right]_1^2 = 3 + \frac{\ln 2}{8}.$$

In passing from line 2 to line 3 we had to be able to spot that $4x^2 + \dfrac{1}{64x^2} + \dfrac{1}{2}$ is a perfect square.

11.2 Surface Area of Revolution

If we take a curve and rotate it about a line we will obtain a curved surface of revolution. For example if we rotate the semicircle in Figure 11.3 about its base through a complete rotation of 2π we will generate a sphere. In general of course we consider not just a semicircle but an arbitrary smooth curve, which we represent in parametric form as $x = x(t)$, $y = y(t)$, $a \le t \le b$. To obtain a formula for such a surface area we divide the curve into small pieces as we did in Section 7.1. In Figure 11.3 we have shown the effect of rotating one such piece of arc. To find the area generated we imagine cutting and unwrapping the section of surface shown. This will give us a piece of "ribbon" approximately rectangular in shape. Its length will be the circumference, $2\pi r$, of the circle generated by rotating a point on the small piece of arc. Its width will be the length ds of the piece of arc, for which we found an approximate formula in Section 11.1. So, using the same notation as in Section 11.1, the small piece of

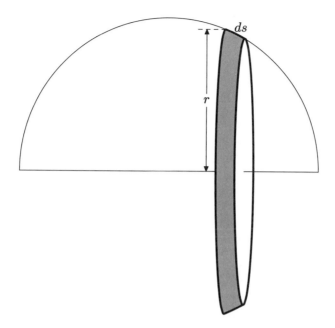

Figure 11.3   Surface of revolution

surface area will be approximately

$$2\pi r.ds \approx 2\pi |y(t)| \sqrt{(x'(t_i)^2 + y'(t_i)^2)} (t_i - t_{i-1}).$$

Note that we have used $|y(t)|$ as the value of r to allow for the fact that $y(t)$ might be negative for some values of t. So the total surface area will be given by

$$S \approx \sum_{i=1}^{n} 2\pi |y(t)| \sqrt{(x'(t_i)^2 + y'(t_i)^2)} (t_i - t_{i-1}).$$

This is an approximating sum for an integral, and so the formula for the surface area of rotation is

$$S = 2\pi \int_a^b |y(t)| \sqrt{(x'(t)^2 + y'(t)^2)} \, dt.$$

If the curve is the graph of a function, $y = f(x)$, $p \le x \le q$, then it can be expressed in the parametric form $x = t$, $y = f(t)$, $p \le t \le q$, as in Section 11.1. The argument in that section shows that in this case

$$S = 2\pi \int_p^q |f(x)| \sqrt{(1 + f'(x)^2)} \, dx = 2\pi \int_p^q |f(x)| \sqrt{1 + \left(\frac{dy}{dx}\right)^2} \, dx.$$

Example 11.3

Find the area of the surface obtained by rotating the curve $y = x^3$, $0 \leq x \leq 1$ about the x-axis.

We use the cartesian formula to obtain

$$
\begin{aligned}
S &= 2\pi \int_0^1 x^3 \sqrt{1 + (3x^2)^2}\, dx = 2\pi \int_0^1 x^3 \sqrt{1 + 9x^4}\, dx \\
&= 2\pi \left[\frac{1}{54} \left(1 + 9x^4\right)^{\frac{3}{2}} \right]_0^1 = \frac{\pi}{27} \left(10^{\frac{3}{2}} - 1\right).
\end{aligned}
$$

Note that because $x^3 \geq 0$ in the interval we do not need the modulus signs.

Example 11.4

Verify the formula for the surface of a sphere of radius a.

In this example we shall use the parametric formula. The sphere can be obtained by rotating the semicircle given by $x^2 + y^2 = a^2$, $y \geq 0$, about the x-axis. We parameterise the semicircle by $x = a\cos t$, $y = a\sin t$, $0 \leq t \leq \pi$. Since $y(t) \geq 0$ for all t in the interval we do not need the modulus signs in the formula, and so we have

$$
\begin{aligned}
S &= 2\pi \int_0^\pi a\sin t \sqrt{a^2 \sin^2 t + a^2 \cos^2 t}\, dt \\
&= 2\pi a^2 \int_0^\pi \sin t\, dt = 2\pi a^2 \left[-\cos t\right]_0^\pi = 4\pi a^2.
\end{aligned}
$$

11.3 Volumes by Slicing

Consider a prism, whose uniform cross section can be any shape, for example triangular, rectangular, or circular (a circular prism is of course a cylinder). Its volume is equal to the cross-sectional area multiplied by its length. We can extend this idea to solids where the cross section is not uniform.

So suppose we have a solid, contained between planes $x = a$ and $x = b$, and that we can calculate the area $A(c)$ of the cross section made by the plane $x = c$. A good way to imagine this is to think of a sliced loaf of bread, where the cross section will vary from one end to the other. We now slice up the solid by means of a sequence of planes

$$
x = x_0(= a), x = x_1, x = x_2, \ldots, x = x_n(= b).
$$

Assuming that the slices are sufficiently thin, and that $A(x)$ is a continuous function of x, the volume of the slice contained between $x = x_{i-1}$ and $x = x_i$ will be approximately $A(x_i)(x_i - x_{i-1})$. The total volume will therefore be approximately the sum of these slice volumes, so

$$V \approx \sum_{i=1}^{n} A(x_i)(x_i - x_{i-1}).$$

This is an approximating sum for an integral, as in Section 7.1, and so we have

$$V = \int_a^b A(x)\, dx.$$

Naturally for this to be useful we have to be able to find $A(x)$ for the solid we are concerned with, and we do this is the next example.

Example 11.5

A tetrahedron is formed by cutting a corner from a cube by means of a plane. Find its volume.

We place the corner of the cube O at the origin, with the sides of the cube meeting at that corner along the coordinate axes. Let the cutting plane meet the coordinate axes at $x = a, y = b, z = c$ respectively. Figure 11.4 shows the tetrahedron with a triangular cross section PQR made by a plane $x = p$. We therefore need to calculate the area of this triangle.

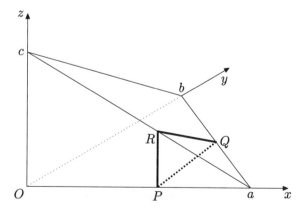

Figure 11.4   Tetrahedron for Example 11.5

Some elementary calculation with similar triangles tells us that

$$PR = \frac{c(a-p)}{a}, \quad PQ = \frac{b(a-p)}{a},$$

so the area of triangle PQR is given by

$$A(p) = \frac{PR.PQ}{2} = \frac{bc(a-p)^2}{2a^2}.$$

Therefore the volume of the tetrahedron is given by

$$V = \int_0^a A(x)\, dx = \int_0^a \frac{bc(a-x)^2}{2a^2}\, dx = \frac{bc}{2a^2}\left[-\frac{(a-x)^3}{3}\right]_0^a = \frac{abc}{6}.$$

11.4 Volumes of Revolution

If we imagine rotating a plane region about a line in that plane then a solid will be generated with circular cross sections. For example if we rotate a rectangular region about one of its edges we obtain a cylinder. A right-angled triangle rotated about one of its shorter edges will generate a cone. A semicircular region rotated about its diameter will generate a sphere. In this section we shall investigate two methods of calculating the volume of a solid of revolution.

11.4.1 The Disc Method

If we rotate a rectangular region about a line parallel to one of its edges which does not intersect the rectangle then we will generate a solid cylinder with a cylindrical hole through its centre. This is a simple example of a general type of solid obtained by rotating the region specified by

$$f(x) \le y \le g(x), \quad a \le x \le b$$

about the line $y = c$, shown in Figure 11.5.

As before we calculate the cross sectional area of the solid and use the integral formula derived above. We subdivide the interval $a \le x \le b$ as in Section 7.1, and use this to divide the region into strips parallel to the y-axis. We have shown one of these in Figure 11.5, together with the solid obtained by rotating this strip about the line $y = c$. It looks like a "washer", i.e., a disc with a smaller disc removed from its centre. The cross-sectional area is therefore

$$\pi R^2 - \pi r^2 = \pi\left((g(x) - c)^2 - (f(x) - c)^2\right).$$

The integral formula tells us that the total volume is given by

$$V = \pi \int_a^b \left((g(x) - c)^2 - (f(x) - c)^2\right)\, dx.$$

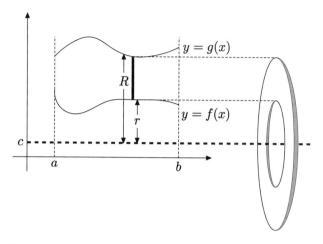

Figure 11.5   Volume of revolution; the disc method

Example 11.6

Find the volume of the solid obtained by rotating the region

$$0 \le y \le \sin x, \quad 0 \le x \le \pi$$

about (a) the x-axis (b) the line $y = -1$.

(a) In this case we have $c = 0$, $f(x) = 0$, $g(x) = \sin x$, and so the volume is given by

$$V = \pi \int_0^\pi \sin^2 x \, dx = \frac{\pi}{2} \left[x - \frac{\sin 2x}{2} \right]_0^\pi = \frac{\pi^2}{2},$$

using the result of Example 7.2 to evaluate the integral.

(b) Here we have $c = -1$ and so the volume is given by

$$
\begin{aligned}
V &= \pi \int_0^\pi \left((\sin x + 1)^2 - (-1)^2 \right) dx \\
&= \pi \int_0^\pi \sin^2 x \, dx + \pi \int_0^\pi 2 \sin x \, dx = \frac{\pi^2}{2} + 4\pi.
\end{aligned}
$$

We can plot both these solids of revolution using the MAPLE `cylinderplot` command. In the first case we use

`with(plots): cylinderplot(sin(z),theta=0..2*Pi,z=0..Pi);`

In the second case we need to recognise that this is equivalent to rotating the region

$$1 \le y \le 1 + \sin x, \quad 0 \le x \le \pi$$

about the x-axis. The command

`with(plots): cylinderplot([1+sin(z),1],theta=0..2*Pi,z=0..Pi);`

will then plot the solid, including the hole through the centre. A MAPLE plot of the second solid is shown in Figure 11.6.

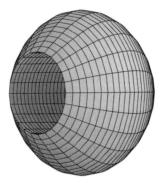

Figure 11.6   MAPLE plot for Example 11.6

11.4.2 The Cylindrical Shell Method

This method applies to the same region

$$f(x) \leq y \leq g(x), \quad a \leq x \leq b.$$

for which we developed the disc method, but this time we rotate the region about the line $x = d$, parallel to the y-axis, as shown in Figure 11.7. In this case the strips produced by subdividing the interval $a \leq x \leq b$ generate cylindrical shells rather than a cross section of the solid. One has to imagine each of these shells fitting inside the previous one to form the solid, like some childrens' toys where plastic beakers fit inside each other, or like the separate parts of Russian dolls.

In Figure 11.7, suppose that the strip is specified by one of the intervals of the subdivision, i.e.,

$$f(x) \leq y \leq g(x), \quad x_{i-1} \leq x \leq x_i.$$

The distance of this strip from the axis of rotation is shown as R, where we have $R = x - d$. (In the diagram d is negative, and so $R = x - d > x$, as the figure suggests.) On the left of the figure is shown the cylindrical shell generated by rotating the strip, and we need to find its volume. If the cylinder is made of some flexible material, we can cut it parallel to the axis of rotation. We can then open it out and we will obtain an approximately rectangular slab. Its height will be the height of the strip, namely $g(x) - f(x)$, its width will be

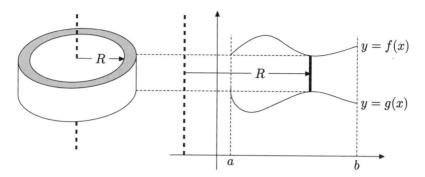

Figure 11.7   Volume of revolution; the shell method

the circumference of the cylinder, namely $2\pi R$, and its thickness will be the width of the strip, namely $x_i - x_{i-1}$. The volume is therefore approximately

$$2\pi R(g(x) - f(x))(x_i - x_{i-1}) = 2\pi(x - d)(g(x) - f(x))(x_i - x_{i-1}).$$

The total volume is therefore approximately

$$\sum_{i=1}^{n} 2\pi(x - d)(g(x) - f(x))(x_i - x_{i-1}),$$

which is an approximating sum for an integral, and hence

$$V = 2\pi \int_a^b (x - d)(g(x) - f(x))\, dx.$$

Example 11.7

Rotate the region in Example 11.6 about the line $x = -\pi$, and find the volume of the solid obtained.

The region is $0 \le y \le \sin x$, $0 \le x \le \pi$, and so using the formula obtained above we have

$$f(x) = 0, \quad g(x) = \sin x, \quad a = 0, \quad b = \pi, \quad d = -\pi.$$

We therefore have

$$V = 2\pi \int_0^\pi (x + \pi)\sin x\, dx = 2\pi^2 \int_0^\pi \sin x\, dx + 2\pi \int_0^\pi x \sin x\, dx = 6\pi^2.$$

Both integrals are straightforward. The second is done by parts, and the first is a basic integral. This integral is encountered frequently, so it is worth learning that its value, which corresponds to the area of the region in this example, is equal to 2. A MAPLE plot of the solid is shown in Figure 11.8.

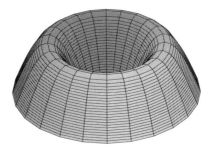

Figure 11.8   MAPLE plot for Example 11.7

Example 11.8

Let R denote the region contained between the two graphs

$$y = x^2, \quad y = \sqrt{x}, \quad 0 \le x \le 1.$$

Find the volume obtained by rotating this region about (a) the x-axis, (b) the y-axis.

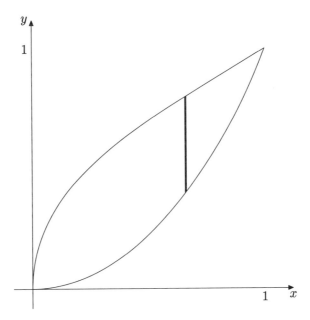

Figure 11.9   Diagram for Example 11.8

The region is shown in Figure 11.9, and we have drawn a strip parallel to the y-axis. If we rotate the strip round the x-axis this will generate a washer, and so we can use the disc method. If we rotate it round the y-axis we will generate a cylinder, so we can use the shell method.

(a) The volume obtained by rotating round the x-axis is given by

$$V = \pi \int_0^1 \left(\sqrt{x}^2 - (x^2)^2 \right) \, dx = \pi \int_0^1 \left(x - x^4 \right) \, dx = \frac{3\pi}{10}.$$

(b) The volume obtained by rotating round the y-axis is given by

$$V = 2\pi \int_0^1 x \left(\sqrt{x} - x^2 \right) \, dx = 2\pi \int_0^1 \left(x^{\frac{3}{2}} - x^3 \right) \, dx = \frac{3\pi}{10}.$$

Because of the symmetry of the region we can see that the two solids will in fact be identical in shape and size. A MAPLE plot, shown in Figure 11.10, can be produced with the command

```
with(plots): cylinderplot([z^2,sqrt(z)],theta=0..2*Pi,z=0..1);
```

Figure 11.10   MAPLE plot for Example 11.8

This plot is much clearer on the MAPLE screen, because it is shaded in colour, and can also be rotated to give varying views of the solid object.

11.5 Density and Mass

Suppose that one of the solids whose volume we have calculated is made of a homogeneous material. Then provided we know the mass of a particular object made of that material, for example a unit cube, then by simple proportion we can calculate the mass of the whole solid. The mass of a unit volume such as a 1cm cube is called the **density**. So far we have not explicitly mentioned units of measurement, but suppose we are measuring length in centimetres and mass in grams. Suppose the solid in Example 11.8 is made of a metal for which the mass of one cubic centimetre is 10 grams, i.e., its density is 10 grams/cm³. Then by proportionality the total mass of the solid will be

$$\frac{3\pi}{10}.10 = 3\pi \ \text{grams}.$$

A problem of more interest arises if a solid is made of a non-homogeneous material. In this case we have to introduce the notion of local density at a point P of the solid. To do this we imagine P to lie at the centre of a small cube of side k cm, for which the total mass is $M(k)$ grams. So the average density of the cube is $\dfrac{M(k)}{k^3}$ grams/cm^3. We now envisage k getting smaller and smaller. If this average density has a limit as k tends to zero we call that limit the **local density** at the point P. We often use the Greek letter ρ for density, and so

$$\rho(P) = \lim_{k \to \infty} \frac{M(k)}{k^3} \text{ grams/cm}^3.$$

In practice we are often given this local density as a function of some coordinates, perhaps $\rho(x, y, z)$.

To calculate the total volume of a solid of variable local density, we imagine it subdivided into small pieces whose volumes V_i we know, for example cubes, over which the local density is approximately constant. In each of these small pieces we choose a point (x_i, y_i, z_i). The mass of a small piece is then approximately $\rho(x_i, y_i, z_i)V_i$, and so the total mass of the solid is given by

$$M = \sum_i \rho(x_i, y_i, z_i)V_i.$$

This looks like an approximating sum for an integral, as we shall find in the examples below. In many problems we have a certain amount of symmetry, in which case the density might depend on only one variable, and we shall be able to obtain an integral in one variable, which we can evaluate.

We have discussed three-dimensional solids above, but we could equally well consider these ideas in one or two dimensions. In one dimension we imagine a length of wire, for which the density is given as mass per unit length (grams/cm). In two dimensions we consider a surface, for which the density is given as mass per unit area (grams/cm^2).

Example 11.9

Suppose we have a length of wire lying along the curve $y = x^3$, $1 \le x \le 2$, whose local linear density at the point (x, y) is equal to x^3 gm/cm. Find the total mass of the length of wire.

We imagine the wire subdivided into small lengths. Using the notation of Section 11.1 the length of a small piece of arc is given approximately by

$$ds = \sqrt{1 + \left(\frac{dy}{dx}\right)^2}\, dx = \sqrt{1 + 9x^4}\, dx.$$

So the mass of this small length is approximately

$$x^3 \sqrt{1 + 9x^4}\, dx.$$

The total mass is the sum of these, which is an approximating sum for an integral. The total mass is therefore given by

$$M = \int_1^2 x^3 \sqrt{1 + 9x^4}\, dx = \left[\frac{2}{3}\frac{1}{9}\frac{1}{4}\left(1 + 9x^4\right)^{\frac{3}{2}}\right]_1^2 = \frac{1}{54}\left(145^{\frac{3}{2}} - 10^{\frac{3}{2}}\right) \text{ gm.}$$

Example 11.10

Suppose we have a disc in the x-y plane of radius 2, centred at the origin, and that its local surface density at the point (x, y) is $(5 + x)\text{gm/cm}^2$. Find the total mass of the disc.

Following the notation of Section 7.1 we subdivide the interval $-2 \le x \le 2$ on the x-axis by the sequence of points

$$-2 = x_0, x_1, x_2, \cdots, x_{n-1}, x_n = 2.$$

We then consider the thin strip, parallel to the y-axis, forming part of the disc, consisting of all the points (x, y) of the disc for which $x_{i-1} \le x \le x_i$. From the equation of the circumference of the disc, $x^2 + y^2 = 4$, we see that the height of this strip is approximately $2\sqrt{4 - x_i^2}$. The area of the strip is then approximately $2\sqrt{4 - x_i^2}\,(x_i - x_{i-1})$ and its mass is therefore approximately

$$2(5 + x_i)\sqrt{4 - x_i^2}\,(x_i - x_{i-1}).$$

The total mass is therefore approximately

$$\sum_{i=1}^n 2(5 + x_i)\sqrt{4 - x_i^2}\,(x_i - x_{i-1}).$$

This is an approximating sum for an integral, and so the mass is given by

$$M = \int_{-2}^2 2(5 + x)\sqrt{4 - x^2}\, dx = 10 \int_{-2}^2 \sqrt{4 - x^2}\, dx + 2 \int_{-2}^2 x\sqrt{4 - x^2}\, dx.$$

The first integral represents the area of a semicircle of radius 2, and the second integral is zero because we are integrating an odd function over an interval symmetric about the origin. Therefore

$$M = 10.\frac{4\pi}{2} = 20\pi \text{ gm.}$$

Example 11.11

A flat plate is in the shape of a plane region bounded by the parabola $y = x^2$ and the line $y = 4$. Its local surface density at the point (x, y) is proportional to the distance of the point from the line $y = -1$, i.e., $\rho(x, y) = 1 + y$. Find the total mass.

In this example we subdivide the appropriate interval on the y-axis by means of the sequence of points

$$0 = y_0, y_1, y_2, \cdots, y_{n-1}, y_n = 4.$$

These serve to subdivide the region into strips parallel to the x-axis. From the equation of the parabola the length of the strip determined by $y_{i-1} \le y \le y_i$ is approximately $2\sqrt{y_i}$. As in the last example we construct approximate formulae for the area and mass of this strip and then sum to determine an approximation for the total mass. We then recognise this as an approximating sum for an integral, and the total mass will therefore be given by

$$M = \int_0^4 2\sqrt{y}\,(1 + y)\,dy = \frac{544}{15}.$$

Readers should draw a diagram and fill in the details of the argument along the lines of Example 11.10. It is important to develop the ability to draw diagrams and to use them to construct appropriate integral formulae for specific problems such as this, rather than trying to manipulate some standard formula, the latter approach often leading to errors.

Example 11.12

Suppose we have a sphere of radius a whose local density at a point P is proportional to the square of distance of P from the centre. Find the total mass of the sphere.

In this example the density is constant at each point of a spherical surface of radius $r \le a$, and so we shall subdivide the sphere into thin spherical shells. We accomplish this by means of a sequence of radii,

$$0 = r_0, r_1, r_2, \cdots, r_{n-1}, r_n = a.$$

The spherical surface determined by r_i has area $4\pi r_i^2$, and so the volume of a corresponding thin spherical shell is approximately $4\pi r_i^2(r_i - r_{i-1})$. Its mass is approximately

$$kr_i^2 4\pi r_i^2(r_i - r_{i-1}) = 4\pi kr_i^4(r_i - r_{i-1}),$$

where k is the constant of proportionality. Summing over the sphere gives an approximating sum for an integral, and so the total mass is given by

$$M = 4\pi k \int_0^a r^4 \, dr = \frac{4\pi ka^5}{5}.$$

11.6 Centre of Mass and Centroid

Imagine that you have a billiard or pool cue and you are trying to find a point along its length where you can balance it on your finger. It will be nearer to the thicker end of the cue, where more of the mass is concentrated. In this section we shall find such balance points for some examples in one, two and three dimensions. The point of balance is called the **centre of mass**. In the case of an object of uniform density then the point of balance depends only on the shape of the object, and it is sometimes called the **centroid**, which we can think of as a geometrical centre of the object.

We can approximate to the mass distribution of the billiard cue by thinking of it as a sequence of small masses concentrated at points along its length. If we are trying to balance it about a point P, the masses on one side of P will tend to pivot the cue in one direction, and those of the other side of P will pivot the cue in the other direction. The effect of each mass will depend upon both how heavy it is and also on how far it is from the point P. The **moment** (or turning moment) of such a point mass about P is defined as md, where m is its mass (in grams or some other unit) and d is the distance of the mass from P. We choose a direction of measurement (usually corresponding to coordinate axes) so that sometimes d will be negative and sometimes d will be positive.

The total moment of the billiard cue, approximated by masses

$$m_1, m_2, \cdots, m_n,$$

about the point P will be given by

$$\sum_{i=1}^{n} m_i d_i,$$

where d_i are the corresponding distances from P. We can see that this looks like an approximating sum for an integral.

The point of balance, or centre of mass, will be the point P for which the total moment is zero, so that the tendency to turn in either direction balances. We can sometimes think of all the mass as being concentrated at that point as far as balance is concerned.

Example 11.13

A billiard cue of length 2m is placed along the positive x-axis, with one end at the origin. We can consider it to be a one-dimensional object, whose local linear density is given by

$$2 + (x - 2)^2 \text{ kg/m}.$$

Find the position of the centre of mass.

Let P denote a point along the cue at distance p from the origin. As always we subdivide the relevant interval on the x-axis by means of a sequence of points

$$0 = x_0, x_1, x_2, \cdots, x_{n-1}, x_n = 2.$$

The mass of that part of the cue between x_{i-1} and x_i is approximately

$$(2 + (x_i - 2)^2)(x_i - x_{i-1}).$$

Its moment about P is

$$(2 + (x_i - 2)^2)(x_i - x_{i-1})(x_i - p).$$

Summing these gives the total moment about P, and gives an approximating sum for the integral

$$\int_0^2 (2 + (x - 2)^2)(x - p) \, dx = \frac{16 - 20p}{3}.$$

The centre of mass occurs where this is zero, i.e., $p = \dfrac{16}{20} = \dfrac{4}{5}$.

Example 11.14

A semicircular flat plate has uniform density. Find its centre of mass.

We choose coordinates as shown in Figure 11.11, where a denotes the radius.

For a two-dimensional example we need to find both coordinates of the centre of mass. In this case by symmetry the centre of mass will clearly lie on the y-axis. To find the y-coordinate of the centre of mass we subdivide the plate into strips as shown in Figure 11.11. The strip shown has length given by $l = 2\sqrt{a^2 - y^2}$. Suppose the density of the plate is ρ. Then the moment of the strip about the x-axis will be

$$y.l.\rho.\delta y = 2\rho y \sqrt{a^2 - y^2} \delta y,$$

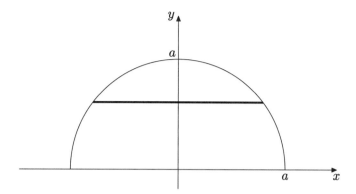

Figure 11.11   Diagram for Example 11.14

where δy is the thickness of the strip. Summing and taking the limit gives the total moment as

$$K = \int_0^a 2\rho y \sqrt{a^2 - y^2} \, dy = 2\rho \left[-\frac{(a^2 - y^2)^{\frac{3}{2}}}{3} \right]_0^a = \frac{2\rho a^3}{3}.$$

Now the total mass of the plate is given by

$$M = \frac{\pi a^2 \rho}{2}.$$

The centre of mass is the place where all the mass may be considered to be concentrated as far as moments are concerned. So if the total mass M were at a distance c from the origin on the y-axis, its moment about the x-axis would be Mc. This must be equal to the total moment of the plate as calculated, and so we have $Mc = K$, i.e.,

$$\frac{\pi a^2 \rho}{2}.c = \frac{2\rho a^3}{3}, \quad \text{giving } c = \frac{4a}{3\pi}.$$

So the centre of mass has coordinates $\left(0, \frac{4a}{3\pi}\right)$.

Example 11.15

Find the centre of mass of an isosceles right-angled triangle whose local surface density at the point P is proportional to the square of the distance of P from one of the shorter edges of the triangle.

Let a denote the length of the shorter edge of the triangle. We choose a subdivision of the triangle into strips parallel to the x-axis, as shown in

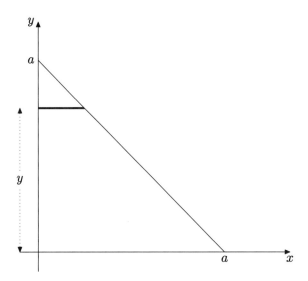

Figure 11.12   Diagram for Example 11.15

Figure 11.12, because the density is uniform along the strip, equal to ky^2, where k is the constant of proportionality.

To find the centre of mass we have to consider both its coordinates, unlike the previous example where we were able to use symmetry to determine one of them. In this case therefore we need to consider moments about both axes. We calculate the various quantities needed as follows, using δy to denote the thickness of the strip.

The length of the strip is $a - y$.

The mass of the strip is $(a - y)ky^2\delta y$.

The moment of the strip about the x-axis is $y(a - y)ky^2\delta y$.

Because the density of the strip is uniform along its length, its centre of mass is half-way along, and so the moment of the strip about the y-axis is $\dfrac{a - y}{2}(a - y)ky^2\delta y$.

We now sum these quantities over the whole triangle and take the limit to give integral expressions as follows.

The total mass is given by

$$M = \int_0^a k(a - y)y^2 \, dy = k\left[\frac{ay^3}{3} - \frac{y^4}{4}\right]_0^a = k\frac{a^4}{12}.$$

The total moment about the x-axis is given by

$$\int_0^a k(a - y)y^3 \, dy = k\left[\frac{ay^4}{4} - \frac{y^5}{5}\right]_0^a = k\frac{a^5}{20}.$$

The total moment about the y-axis is given by

$$\int_0^a \frac{k}{2}(a-y)^2 y^2 \, dy = \frac{k}{2}\left[\frac{a^2 y^3}{3} - 2\frac{ay^4}{4} + \frac{y^5}{5}\right]_0^a = k\frac{a^5}{60}.$$

Now suppose that the centre of mass has coordinates (X, Y). Then

$$MX = k\frac{a^5}{20}, \quad \text{i.e.,} \quad k\frac{a^4}{12}X = k\frac{a^5}{20}, \quad \text{so} \quad X = \frac{3a}{5}.$$

$$MY = k\frac{a^5}{60}, \quad \text{i.e.,} \quad k\frac{a^4}{12}Y = k\frac{a^5}{60}, \quad \text{so} \quad Y = \frac{a}{5}.$$

Therefore the centre of mass is located at the point $\left(\dfrac{3a}{5}, \dfrac{a}{5}\right)$.

Example 11.16

A solid paraboloid of revolution is obtained by rotating the part of the parabola $z = x^2$ for which $0 \leq z \leq 4$ about the z-axis. Its density is $\rho(x, y, z) = k\sqrt{z}$. Find the centre of mass of the solid.

This is a three-dimensional example, but we can use the symmetry of the problem to say that the centre of mass will be on the axis of rotation of the solid, namely the z-axis. We therefore have only to find its z coordinate, which we denote by Z.

Part of the skill of solving problems such as this consists of drawing an appropriate diagram, and in this case, as in some previous examples, we shall leave the reader to construct a diagram from the descriptions given below, rather than providing one.

We subdivide the solid into discs obtained by slicing the solid perpendicular to the z-axis, i.e., parallel to the x-y plane. From the equation of the parabola, $z = x^2$, we can see that the radius of such a disc at distance z from the origin will be $\sqrt{z}$. Its area will be πz, and so if δz denotes its thickness its mass will be $\pi z.k\sqrt{z}\,\delta z$. The centre of mass of this disc is at the geometrical centre, on the z-axis, by symmetry, and so the moment of the disc about the x-axis is $z.\pi z.k\sqrt{z}\,\delta z$.

The total mass is given by

$$\int_0^4 k\pi z^{\frac{3}{2}} \, dz = k\pi\left[\frac{2}{5}z^{\frac{5}{2}}\right]_0^4 = \frac{2k\pi}{5}.32.$$

The total moment is given by

$$\int_0^4 k\pi z^{\frac{5}{2}} \, dz = k\pi\left[\frac{2}{7}z^{\frac{7}{2}}\right]_0^4 = \frac{2k\pi}{7}.128.$$

The coordinate Z therefore satisfies

$$\frac{2k\pi}{5}.32.Z = \frac{2k\pi}{7}.128.$$

Therefore the centre of mass is located at the point $\left(0,0,\frac{20}{7}\right)$.

EXERCISES

11.1. Find the length of the curve given by $x = t^4$, $y = t^6$, $1 \le t \le 3$.

11.2. Find the length of the curve given by $y = \frac{x^3}{2} + \frac{1}{6x}$, $2 \le x \le 3$.

11.3. Find the length of the curve given by $y = x^2 - \frac{\ln x}{8}$, $1 \le x \le 2$.

11.4. Find the length of the curve given by $y = \cosh x$, $0 \le x \le \alpha$. Determine the value of α for which the length of the curve is 10. Find an approximation correct to four decimal places using your calculator.

11.5. Calculate the area of the surface obtained by rotating the curve

$$x = \frac{t^3}{3}, \ y = \frac{t^2}{2}, \ 0 \le t \le 1$$

about the x-axis.

11.6. The ellipse whose parametric equations are $x = 2\cos t$, $y = \sin t$, is rotated around the x-axis. Find the surface area of the resulting solid.

11.7. Adapt the argument at the beginning of Section 11.2 to show that the surface area obtained by rotating the curve specified by

$$x = x(t), \ y = y(t), \ a \le t \le b$$

about the y-axis is given by the integral formula

$$S = 2\pi \int_a^b |x(t)|\sqrt{(x'(t)^2 + y'(t)^2)} \, dt.$$

The surface of a parabolic mirror is generated by rotating the curve $y = 0.1x^2$, $0 \le x \le 1$ about the y-axis. Find its surface area, and compare it with the area of a circular plane mirror of radius 1.

11.8. Find the volume of the conical solid formed by joining all the points of the ellipse with equation $\dfrac{x^2}{4} + \dfrac{y^2}{9} = 1$ in the x-y plane to the point $(0,0,4)$ on the z-axis.

[You can use the fact that the area of the ellipse with equation $\dfrac{x^2}{a^2} + \dfrac{y^2}{b^2} = 1$ is πab.]

11.9. Find the volume of revolution of the solids obtained by rotating the region specified by

$$1 \le y \le 1 + \cos x, \quad -\frac{\pi}{2} \le x \le \frac{\pi}{2}$$

about the lines (i) $y = -1$, (ii) $x = -\pi/2$.

11.10. Find the volume of the solids obtained by rotating a circular disc of radius a about (i) a tangent line, (ii) a line distance $b > a$ from the centre of the disc.

11.11. A solid is formed by rotating the region bounded by the curves

$$y = \cosh x, \ y = \sinh x, \ x = 0, \ x = 1$$

about the y-axis. Calculate its volume.

11.12. A length of wire lies along the x-axis ($0 \le x \le 1$). Its linear density is given by $\rho(x) = e^x$. Find the mass and centre of mass of the wire.

11.13. Find the centroid of a flat plate in the shape of a quarter of a circle.

11.14. A flat plate is in the shape of a quadrant of a unit circle. Its surface density is proportional to the distance from one of its straight edges. Find the mass and centre of mass.

11.15. A plane region consists of an equilateral triangle, joined to a semi-circle having one of the sides of the triangle as its diameter. Find the centroid of this region.

11.16. A surface is in the shape of a right circular cone, including its circular base. The base has radius 1 and the height is 2. The surface density is proportional to the distance from the vertex of the cone. Find the mass and centre of mass of the curved surface, excluding the base, and also of the complete surface, including the base.

11.17. Consider a solid cone with the same dimensions as that in Exercise 11.16, and with density equal to $(2+x)$, where x is the distance from the circular base. Find the mass and centre of mass.

11.18. Find the centroid of a solid hemisphere.

Answers to Exercises

Chapter 1

1.1 (a) $x \neq$ odd multiples of $\pi/4$;   (b) All real x;
  (c) $x < 1$;   (d) $-1 < x < 1$;
  (e) $x \neq 2$, $x \neq 3$;   (f) $x > 0$;
  (g) $x \leq -1$ or $x \geq 4$;   (h) $x \leq 1$, $x \neq 0$;
  (i) $x < -2$ or $x > 2$ and $x \neq \sqrt{5}$;   (j) $x \neq$ multiples of $\pi/2$;
  (k) $x \neq 0$, $x \neq$ odd multiples of $\pi/2$;   (l) $x \neq 0$.

1.2
Expression	Domain	Range
$\sqrt{x}$	$x \geq 0$	$y \geq 0$
$\sqrt{-x}$	$x \leq 0$	$y \geq 0$
$-\sqrt{x}$	$x \geq 0$	$y \leq 0$
$-\sqrt{-x}$	$x \leq 0$	$y \leq 0$

1.3 Maximal domain: $x \neq 5$. Range: $y \neq 2$.

1.4 Ranges are as follows.

    (a) All real y;   (b) All real y;   (c) All real y;
    (d) $y \leq 0$;   (e) $y > 0$;   (f) $y \geq 0$;
    (g) $y \geq 0$;   (h) $y \geq 1$;   (i) $y \neq 0$;
    (j) $y \leq -1$, $y \geq 1$;   (k) $y \neq 0$;   (l) $y > 0$, $y < -1$.

1.5 (a) even;   (b) neither;   (c) odd;   (d) even;   (e) even;
    (f) even;   (g) odd;   (h) neither;   (i) neither.

1.6 $f(-x)g(-x) = (-f(x))(-g(x)) = f(x)g(x)$.

$E_1(-x)E_2(-x) = E_1(x)E_2(x)$.

$E(-x)O(-x) = E(x)(-O(x)) = -E(x)O(x)$.

1.7 $f \circ f(x) = f(f(x)) = \dfrac{3}{3/x} = x, \ (x \neq 0).$

$f \circ g(x) = f(g(x)) = 3\dfrac{2+x}{2-x}, \ x \neq 2, x \neq -2.$

$g \circ f(x) - g(f(x)) = \dfrac{2 - f(x)}{2 + f(x)} = \dfrac{2 - (3/x)}{2 + (3/x)} = \dfrac{2x - 3}{2x + 3}, \ \left(x \neq -\dfrac{3}{2}\right).$

$g \circ g(x) = g(g(x)) = \dfrac{2 - \frac{2-x}{2+x}}{2 + \frac{2-x}{2+x}} = \dfrac{2 + 3x}{6 + x}, \ x \neq -2, x \neq -6.$

1.8 The graphs needed are e^x and $\sin x$. The composite graph can be checked using MAPLE or a graphical calculator.

1.9 $e^{g(x)} = 3x - 4, \ $ so $ \ g(x) = \ln(3x - 4).$

1.10 $f(x) = \dfrac{4}{(x-1)^2}.$

1.11 (a) $1 + \dfrac{4x^2 + 2}{x^4 - 2x^2 - 1};$ (b) $x^2 - 2x + 4 - \dfrac{7}{x+2};$

(c) $\dfrac{x^3}{2} - \dfrac{7x^2}{4} + \dfrac{43x}{8} - \dfrac{215}{16} + \dfrac{963}{16(2x+5)};$

(d) $x^3 - x^2 - 3 + \dfrac{4x - 10}{x^2 + x - 3};$

(e) $x^4 + x^3 + 3x^2 + 5x + 11 + \dfrac{21x + 21}{x^2 - x - 2};$

(f) $x^2 - 3x + 8 - \dfrac{8x + 13x^2 - 24}{x^3 + 2x^2 - 3}.$

1.12 (a) $(x+1)(x-2)(x+2);$ (b) $(y-1)(y+1)^2;$
      (c) $(z+3)(z+4)(z-5);$ (d) $(c-1)^2(c+1)^3;$
      (e) $(t-1)(t+1)(t^2 + t + 1)(t^2 - t + 1);$
      (f) $(u-1)(u+3)(u+4)(u-5).$

1.13 $g(-2) = 0, \ g'(-2) = g''(-2) = g'''(-2) = 0, g^{(4)}(-2) = -24.$

So -2 is a root of multiplicity 4.

1.14 (a) $x = \dfrac{1}{2}, x = \dfrac{9}{2};$ (b) $x = 1 + \sqrt{7}, 1 - \sqrt{7}, 1 + \sqrt{5}, 1 - \sqrt{5};$

(c) $x = 2, x = \sqrt[3]{-6};$ (d) $-\dfrac{7}{2} < x < -\dfrac{1}{2};$

(e) $-1 \leq x \leq \dfrac{5 - \sqrt{33}}{4} \ $ or $ \ \dfrac{5 + \sqrt{33}}{4} \leq x \leq \dfrac{7}{2};$

(f) $x \leq \dfrac{5 - \sqrt{33}}{2} \ $ or $ \ x \geq \dfrac{5 + \sqrt{33}}{2} \ $ or $ \ \dfrac{5 - \sqrt{17}}{2} \leq x \leq \dfrac{5 + \sqrt{17}}{2}.$

1.15 The graphs without the modulus can be plotted, and those parts below the x-axis reflected in the x-axis. They can also be plotted using MAPLE.

1.16 $\cos 5x = 16 \cos^5 x - 20 \cos^3 x + 5 \cos x$.

1.17 $\operatorname{cosec}(x + y) = \dfrac{\operatorname{cosec} x \sec x \operatorname{cosec} y \sec y}{\sec x \operatorname{cosec} y + \operatorname{cosec} x \sec y}$.

1.18 Let $A = \ln p$, $B = \ln q$; $p = e^A$, $q = e^B$; $\dfrac{p}{q} = e^{A-B}$;

$A - B = \ln\left(\dfrac{p}{q}\right)$; $\ln p - \ln q = \ln\left(\dfrac{p}{q}\right)$

1.19 (a) 1; (b) $-\frac{3}{2}$; (c) 0.

1.20 $\sinh 3x = 4 \sinh^3 x + 3 \sinh x$.

1.21 These graphs can be sketched using similar reasoning to that given for the graphs of the reciprocals of the trigonometric functions. They can be checked using MAPLE.

1.22 (a) $\dfrac{1}{2}\left(x + \dfrac{1}{x}\right)$; (b) $\dfrac{x^2 + 1}{x^2 - 1}$; (c) $\dfrac{1}{x^2}$.

1.23 (a) The domain is $x \neq \dfrac{1}{3}$. $f^{-1}(x) = \dfrac{x}{3x - 2}$, $x \neq \dfrac{2}{3}$.

(b) The domain is the set of all real x. $f^{-1}(x) = \dfrac{\sqrt{3}x}{\sqrt{4 - x^2}}$, $(-2 < x < 2)$.

1.24 (a) yes; (b) no; (c) yes; (d) yes; (e) no; (f) yes; (g) yes; (h) yes.

1.25 If $a \leq b$, then $f(a) \leq f(b)$ and $g(a) \leq g(b)$. Using the rules for inequalities then gives $f(a) + g(a) \leq f(b) + g(b)$, so $f + g$ is an increasing functions.

Examples: $f(x) = x^3$ and $g(x) = x$ both increasing, $x^3 - x$ is not.

$f(x) = x$ and $g(x) = x - 1$ both increasing, $x(x - 1)$ is not.

$f(x) = x$ and $g(x) = x^2 + 1$ both increasing, $\dfrac{x}{x^2 + 1}$ is not.

1.26 $y = 5 - 12x - 2x^2 = -2(x+3)^2 + 23$. $x = -3 \pm \sqrt{\dfrac{23 - y}{2}}$. Now $x \geq -3$,

so $f^{-1}(x) = -3 + \sqrt{\dfrac{23 - x}{2}}$.

1.27 (a) $\cos(\sin^{-1} x) = \sqrt{1 - x^2}$; (b) $\sin(\tan^{-1} x) = \dfrac{x}{\sqrt{1 + x^2}}$;

(c) $\tan(\sec^{-1} x) = \sqrt{x^2 - 1}$.

1.28 Let $x = \dfrac{1}{2}$. $\dfrac{\sin^{-1}\frac{1}{2}}{\cos^{-1}\frac{1}{2}} = \dfrac{\pi/3}{\pi/6} = \dfrac{1}{2}$. $\tan^{-1}\dfrac{1}{2} \approx 0.4636$.

1.29 MAPLE uses `arcsinh`, `arccosh` and `arctanh` for the inverse functions.

1.30 Let $y = \sinh x = \dfrac{e^x - e^{-x}}{2}$. So $(e^x)^2 - 2ye^x - 1 = 0$, with solutions $e^x = y \pm \sqrt{y^2 + 1}$. $e^x > 0$, and so $\sinh^{-1} x = \ln\left(x \pm \sqrt{x^2 + 1}\right)$.

Let $y = \tanh x = \dfrac{e^x - e^{-x}}{e^x + e^{-x}} = \dfrac{e^{2x} - 1}{e^{2x} + 1}$. Rearranging this equation gives

$e^{2x} = \dfrac{1 + y}{1 - y}$, and so $x = \dfrac{1}{2}\ln\left(\dfrac{1 + y}{1 - y}\right)$. Hence $\tanh^{-1} x = \dfrac{1}{2}\ln\left(\dfrac{1 + x}{1 - x}\right)$.

Chapter 2

2.1 $f(x) \to 0$ as $x \to \pm\infty$, ∞ as $x \to 2^-$, $-\infty$ as $x \to 2^+$, $-\infty$ as $x \to 3^-$, ∞ as $x \to 3^+$.

2.2

	$f(x) \to$	as $x \to$		$f(x) \to$	as $x \to$
(a)	∞	∞	(k)	$-\infty$	-2^+
	∞	$-\infty$		$-\infty$	4^+
(b)	∞	$-\infty$		∞	4^-
	$-\infty$	∞	(l)	∞	$-\frac{\pi}{4} + n\pi$
(c)	$-\infty$	$-\infty$	(m)	0	$-\infty$
	∞	∞		∞	∞
(d)	0	$\pm\infty$		∞	0
	$-\infty$	-2^-	(n)	$-\infty$	$-\infty$
	∞	-2^+		0	∞
	∞	1^-	(o)	∞	$-\infty$
	$-\infty$	1^+		∞	∞
	$-\infty$	4^-		$-\infty$	-2
	∞	4^+		$-\infty$	2
(e)	∞	$\pm\infty$	(p)	0	$-\infty$
(f)	0	$\pm\infty$		0	∞
(g)	$\frac{1}{3}$	$\pm\infty$		∞	-2
(h)	3	$\pm\infty$		$-\infty$	0^-
	∞	-1^-		∞	0^+
	$-\infty$	-1^+		$-\infty$	2
	$-\infty$	2^-	(q)	$\ln 2$	$\pm\infty$
	∞	2^+		no limit	0

	$f(x) \to$	as $x \to$		$f(x) \to$	as $x \to$
(i)	∞	∞	(r)	$-\infty$	$\pm\infty$
	$-\infty$	$-\infty$		∞	0
	$-\infty$	$\frac{5}{2}^{-}$	(s)	$\frac{\pi}{2}$	∞
	∞	$\frac{5}{2}^{+}$		$-\frac{\pi}{2}$	$-\infty$
(j)	0	$\pm\infty$	(t)	∞	1^{-}
(k)	0	$\pm\infty$		$-\infty$	-1^{+}
	∞	-2^{-}			

2.3 (a) $\exp\left(\dfrac{1}{x}\right) \to \begin{cases} \infty & \text{as} & x \to 0^{+}, \\ 0 & \text{as} & x \to 0^{-}; \end{cases}$

(b) $\sqrt{\text{floor}\sqrt{x}} \to \begin{cases} \sqrt{3} & \text{as} & x \to 9^{+}, \\ \sqrt{2} & \text{as} & x \to 9^{-}; \end{cases}$

(c) $\dfrac{|x|}{x} \to \begin{cases} 1 & \text{as} & x \to 0^{+}, \\ -1 & \text{as} & x \to 0^{-}; \end{cases}$ (d) $\dfrac{|\sin x|}{\sin x} \to \begin{cases} -1 & \text{as} & x \to \pi^{+}, \\ 1 & \text{as} & x \to \pi^{-}; \end{cases}$

(e) $\dfrac{\sqrt{x^2 - 2x + 1}}{x - 1} \to \begin{cases} 1 & \text{as} & x \to 1^{+}, \\ -1 & \text{as} & x \to 1^{-}; \end{cases}$

(f) $\dfrac{\tan x}{|x|} \to \begin{cases} 1 & \text{as} & x \to 0^{+}, \\ -1 & \text{as} & x \to 0^{-}. \end{cases}$

2.4 (a) 4; (b) $2\pi - 3$; (c) $\sqrt{2}$; (d) 12; (e) 1; (f) $\dfrac{1}{6}$; (g) -2; (h) 1; (i) 0; (j) 0.

2.5 (a) $-|x| \le |x| \sin\left(\dfrac{1}{x}\right) \le |x|$. $|x| \to 0$ as $x \to 0$;

(b) $-e^{-x} \le e^{-x} \cos x \le e^{-x}$. $e^{-x} \to 0$ as $x \to \infty$;

(c) $-e^{x} \le e^{x} \sin(x^2 + 1) \le e^{x}$. $e^{x} \to 0$ as $x \to -\infty$;

(d) $\tanh x - 1 \le (1 - \tanh x)\cos x \le 1 - \tanh x$. $\tanh x \to 1$ as $x \to \infty$;

(e) $-|x - 1| \le |x - 1| \cos\left(\dfrac{1}{x - 1}\right) \le |x - 1|$. $|x - 1| \to 0$ as $x \to 1$;

(f) $\exp(\sin x - x) = e^{\sin x} e^{-x}$.

$\quad e^{-1} e^{-x} \le e^{\sin x} e^{-x} \le e^{1} e^{-x}$. $e^{-x} \to 0$ as $x \to \infty$.

2.6 $A \sin x \le f(x) \sin x \le B \sin x$. So by squeezing $\lim\limits_{x \to 0} f(x) \sin x = 0$.

2.7

	−2	−1	0	1	2
$x+1$	−	−	+	+	+	+
$(x+2)^2$	+	+	+	+	+	+
x	−	−	−	+	+	+
$x-1$	−	−	−	−	+	+
$(x-2)^2$	+	+	+	+	+	+
$f(x)$	−	−	+	−	+	+

There are vertical asymptotes at $x = -2, x = 0, x = 1$.

$f(x) \to -\infty$ as $x \to -2$, $f(x) \to \infty$ as $x \to 0^-$, $f(x) \to -\infty$ as $x \to 0^+$,

$f(x) \to -\infty$ as $x \to 1^-$, $f(x) \to \infty$ as $x \to 1^+$.

2.8 (a) 0; (b) 4; (c) 3; (d) 2; (e) $-\dfrac{1}{4}$; (f) 4; (g) -8; (h) -2;

(i) $\dfrac{1}{2}$; (j) 0; (k) ∞; (l) -1; (m) $-\sqrt{2}$; (n) $-\dfrac{2}{3}$; (o) $\dfrac{1}{2}$; (p) $\dfrac{1}{2}$.

2.9 (a) Let $t = e^x$. Then $t \to 1$ as $x \to 0$.

$$\frac{e^{2x} - 1}{e^x - 1} = \frac{t^2 - 1}{t - 1} = t + 1 \to 2 \text{ as } t \to 1.$$

(b) Let $t = \ln x$. Then $t \to 0$ as $x \to 1$.

$$\frac{\sin(\ln x)}{\ln x} = \frac{\sin t}{t} \to 1 \text{ as } t \to 0.$$

(c) Let $t = 2\sqrt{x}$. Then $t \to 0$ as $x \to 0$.

$$\frac{\sqrt{2x}}{\sin((2\sqrt{x}))} = \frac{2\sqrt{x}}{\sqrt{2}\sin(2\sqrt{x})} = \frac{t}{\sqrt{2}\sin t} \to \frac{1}{\sqrt{2}} \text{ as } t \to 0.$$

(d) Let $t = \ln x$. Then $t \to 1$ as $x \to e$.

$$\frac{(\ln x)^2 - 1}{\ln x - 1} = \frac{t^2 - 1}{t - 1} = t + 1 \to 2 \text{ as } t \to 1.$$

(e) Let $t = \sin^{-1} x$. Then $x = \sin t$ and $t \to 0$ as $x \to 0$.

$$\frac{x}{\sin^{-1} x} = \frac{\sin t}{t} \to 1 \text{ as } t \to 0.$$

(f) Let $t = e^x$. Then $t \to 0$ as $x \to -\infty$.

$$\frac{\sin^{-1}(e^x)}{e^x} = \frac{\sin^{-1} t}{t} \to 1 \text{ as } t \to 0, \text{ using part (e).}$$

2.10 (a) $\dfrac{1}{2}$; (b) -1; (c) 0; (d) 0; (e) $\dfrac{a^2}{b^2}$; (f) $\dfrac{a^2}{2}$; (g) -3; (h) $\dfrac{1}{120}$;

(i) 1; (j) 1; (k) e^2; (l) e.

2.11 The following are very simple examples.

(a) $f(x) = x$, $g(x) = x$; (b) $f(x) = 2x$, $g(x) = x$;

(c) $f(x) = x$, $g(x) = 2x$; (d) $f(x) = x + 3$, $g(x) = x$.

2.12 If $k > 0$ $\dfrac{|x + k|}{x^2 - k^2} \to \infty$ as $x \to k^+$, and $\dfrac{|x + k|}{x^2 - k^2} \to -\infty$ as $x \to k^-$.

If $k < 0$ $\dfrac{|x + k|}{x^2 - k^2} \to -\infty$ as $x \to k^+$, and $\dfrac{|x + k|}{x^2 - k^2} \to \infty$ as $x \to k^-$.

For $x > -k$, $|x + k| = x + k$, so $\dfrac{|x + k|}{x^2 - k^2} = \dfrac{1}{x - k} \to -\dfrac{1}{2k}$ as $x \to -k^+$.

For $x < -k$, $|x + k| = -(x + k)$, so $\dfrac{|x + k|}{x^2 - k^2} = \dfrac{-1}{x - k} \to \dfrac{1}{2k}$ as $x \to -k^-$.

Chapter 3

3.1 (a) $\dfrac{(x + h)^3 - x^3}{h} = 3x^2 + 3xh + h^2 \to 3x^2$ as $h \to 0$;

(b) $\dfrac{1}{h}\left(\dfrac{1}{x + h} - \dfrac{1}{x}\right) = \dfrac{-1}{(x + h)x} \to -\dfrac{1}{x^2}$ as $h \to 0$;

(c) $\dfrac{\cos(x + h) - \cos x}{h} = -\sin\left(x + \dfrac{h}{2}\right)\dfrac{\sin\left(\frac{h}{2}\right)}{\frac{h}{2}} \to -\sin x$ as $h \to 0$;

(d) $\dfrac{\tan(x + h) - \tan x}{h} = \dfrac{\sin h}{h}\dfrac{1}{\cos(x + h)\cos x} \to \sec^2 x$ as $h \to 0$;

(e) $\dfrac{e^{x+h} - e^x}{h} = e^x\left(\dfrac{e^h - 1}{h}\right) \to e^x$ as $h \to 0$.

3.2 (a) $6x^{-1/4}$; (b) $\cosh x$; (c) $e^v \cos v + e^v \sin v$; (d) $x^2 \sec^2 x + 2x \tan x$;

(e) $t \cos t$; (f) $\operatorname{sech}^2 x$; (g) $\dfrac{-5}{(2x - 3)^2}$; (h) $\dfrac{-2t^2 - 2t - 2}{(t^2 - 1)^2}$;

(i) $\dfrac{-4x - 2}{3x^{5/3}}$; (j) $\dfrac{-2 - \sin x}{(1 + 2\sin x)^2}$; (k) $\dfrac{e^w(1 - \tan w + \sec^2 w)}{(1 - \tan w)^2}$;

(l) $\dfrac{e^x\left((2x^3 - 5x^2 - 2x)\ln x + x + 2x^2\right)}{(x^2 + 2x^3)^2}$.

3.3 (a) $-\dfrac{\sin(\sqrt{x})}{2\sqrt{x}}$; (b) $-\sinh(\cos t)\sin t$, (c) $-2^{-x}\ln 2$;

(d) $\dfrac{1}{x \ln x \ln(\ln x)}$; (e) $\dfrac{\sqrt{1 + s^{2/3}}}{\sqrt[3]{s}}$; (f) $3t(3 - 2t^2)^{-7/4}$;

(g) $-\dfrac{\sec^2\left(\frac{1}{x}\right)}{x^2}$; (h) $\dfrac{v\cos(v^2)}{\sqrt{\sin(v^2)}}$; (i) $-6\sin 3x\cos(2\cos 3x)$;

(j) $3^{(3^x)}3^x(\ln 3)^2$; (k) $-\dfrac{\sin(\ln x)}{x}$; (l) $\dfrac{1}{3t(\ln t)^{2/3}}$.

3.4 (a) $\dfrac{\sin x + x\cos x}{x\sin x}$; (b) $\cos\left(\dfrac{x}{\cos x}\right)\dfrac{\cos x + x\sin x}{\cos^2 x}$;

(c) $\dfrac{1+e^x}{2\sqrt{x+e^x}}$; (d) $\dfrac{2x^{3/2}}{3(1+x^2)^{2/3}} + \dfrac{\sqrt[3]{1+x^2}}{2\sqrt{x}}$;

(e) $\sinh(x\ln x)(1+\ln x)$; (f) $2x\cos(2x^2)$; (g) $f'(x)\equiv 0$;

(h) $2a^2 x\sec^2\left(a^2(1+x^2)\right)$; (i) $(x\cos x + \sin x)2^{x\sin x}\ln 2$;

(j) $ab(\cos ax + \cos bx)$; (k) $b^2(x^2\ln x)^{b^2-1}(x+2x\ln x)$;

(l) $-4\dfrac{cx}{(cx^2+d)^2}\tan\left(\dfrac{1}{cx^2+d}\right)\sec^2\left(\dfrac{1}{cx^2+d}\right)$.

3.5 A general formula is as follows. It can be proved by induction.

$$
\begin{aligned}
f^{(2n)}(x) &= (-1)^n 2^{2n}\sin(2x+5),\\
f^{(2n+1)}(x) &= (-1)^n 2^{2n+1}\cos(2x+5).
\end{aligned}
$$

Chapter 4

4.1 Note that because of the relationship between x and y one can obtain apparently different, but nevertheless equivalent, expressions.

(a) $\dfrac{dy}{dx} = \dfrac{2y+1}{3-2x}$; (b) $\dfrac{dy}{dx} = -\dfrac{x+1}{3(y-1)^2}$; (c) $\dfrac{dy}{dx} - \dfrac{y}{x}$;

(d) $\dfrac{dy}{dx} = \dfrac{xy\ln y - y^2}{xy\ln x - x^2}$; (e) $\dfrac{dy}{dx} = \dfrac{3y^2 - 2x - y^3}{3y^2 x - 6xy - 2y}$;

(f) $\dfrac{dy}{dx} = -\dfrac{3xy+2y^2}{3xy+2x^2}$; (g) $\dfrac{dy}{dx} = \dfrac{y}{x}$; (h) $\dfrac{dy}{dx} = \dfrac{2xy+1-2y^2}{4xy+3y^2-x^2+1}$;

(i) $\dfrac{dy}{dx} = \dfrac{1-2xy\sin(yx^2) - y^2\cos(xy^2)}{2xy\cos(xy^2) + x^2\sin(xy^2)}$; (j) $\dfrac{dy}{dx} = \dfrac{xye^x + ye^x - ye^y}{xe^y - xe^x + xye^y}$.

4.2 (a) $\dfrac{dy}{dx} = \dfrac{2-y}{3+x}$, $\dfrac{d^2y}{dx^2} = \dfrac{2(y-2)}{(3+x)^2}$;

(b) $\dfrac{dy}{dx} = \dfrac{\cos x - \sin y}{x\cos y}$,

$$
\dfrac{d^2y}{dx^2} = \dfrac{\sin y(\cos x - \sin y)^2}{x^2\cos^3 y} - \dfrac{2(\cos x - \sin y)}{x^2\cos y} - \dfrac{\sin x}{x\cos y}.
$$

4.3 $\dfrac{dy}{dx} = -\dfrac{x}{2y}$, $\dfrac{d^2y}{dx^2} = -\dfrac{1}{y^3}$.

4.4 $3\dfrac{dy}{dx} = \dfrac{3x^2 - y}{x + 6y}$. The gradient at $(2, -2)$ is $-\dfrac{7}{5}$.

The equation of the tangent is $y + 2 = -\dfrac{7}{5}(x - 2)$, i.e., $5y + 7x = 4$.

4.5 $x\sin(xy) = x^2 - 1$. $\sin(xy) + x\cos(xy)\left(y + x\dfrac{dy}{dx}\right) = 2x$. Substituting

$(x, y) = (1, 0)$ gives $\dfrac{dy}{dx} = 2$. The equation of the tangent is therefore $y = 2x - 2$. The tangent at $(-1, 0)$ has equation $y = -2x - 2$.

4.6 (a) $\dfrac{dy}{dx} = x^x(1 + \ln x)$; (b) $\dfrac{dy}{dx} = -x^{-x}(1 + \ln x)$;

(c) $\dfrac{dy}{dx} = -(-x)^{-x}(1 + \ln(-x))$; (d) $\dfrac{dy}{dx} = (\sin x)^{\sin x}\cos x(1 + \ln(\sin x))$;

(e) $\dfrac{dy}{dx} = (e^x)^{\ln x}(1 + \ln x)$; (f) $\dfrac{dy}{dx} = (\ln x)^x\left(\ln(\ln x) + \dfrac{1}{\ln x}\right)$;

(g) $\dfrac{dy}{dx} = 2(\tan x)^{2x}\left(\ln(\tan x) + \dfrac{x}{\sin 2x}\right)$; (h) $\dfrac{dy}{dx} = 2^{x+x^2}(1 + 2x)\ln 2$;

(i) $\dfrac{dy}{dx} = \dfrac{1}{2}\sqrt{e^x \sin x}(1 + \cot x)$;

(j) $\dfrac{dy}{dx} = \sqrt{(x-1)^2 e^{-x}\cos x}\left(\dfrac{1}{x-1} - \dfrac{1}{2} - \dfrac{\tan x}{2}\right)$.

4.7 (a) $\dfrac{dy}{dx} = 2t^2 - t$, $\dfrac{d^2y}{dx^2} = (4t - 1)t$;

(b) $\dfrac{dy}{dx} = \dfrac{1 - 2t}{1 + 2t}$, $\dfrac{d^2y}{dx^2} = -\dfrac{4}{(1 + 2t)^3}$;

(c) $\dfrac{dy}{dx} = \dfrac{2}{1 + \ln t}$, $\dfrac{d^2y}{dx^2} = -\dfrac{2}{t(1 + \ln t)^3}$;

(d) $\dfrac{dy}{dx} = \dfrac{e^t}{2t}$, $\dfrac{d^2y}{dx^2} = \dfrac{e^t(t - 1)}{4t^3}$;

(e) $\dfrac{dy}{dx} = \dfrac{1}{3t\sqrt{t^2 + 1}}$, $\dfrac{d^2y}{dx^2} = -\dfrac{2t^2 + 1}{9t^4(t^2 + 1)^{3/2}}$;

(f) $\dfrac{dy}{dx} = -\dfrac{\sin t}{2t\cos(t^2)}$, $\dfrac{d^2y}{dx^2}$

$= \dfrac{\sin t\,\cos(t^2) - 2t^2\sin t\,\sin(t^2) - t\cos t\,\cos(t^2)}{4t^3\,\cos^3(t^2)}$;

(g) $\dfrac{dy}{dx} = e^t \cos^2 t, \quad \dfrac{d^2y}{dx^2} = (e^t \cos^2 t - e^t.2\cos t \sin t)\cos^2 t;$

(h) $\dfrac{dy}{dx} = -\dfrac{\cos^2 t}{\sin t}, \quad \dfrac{d^2y}{dx^2} = \dfrac{\cos^2 t(\cos^2 t - 2)}{\sin^3 t};$

(i) $\dfrac{dy}{dx} = -\dfrac{e^t + te^t}{\sin t}, \quad \dfrac{d^2y}{dx^2} = \dfrac{e^t((2+t)\sin t) - (1+t)\cos t}{\sin^3 t};$

(j) $\dfrac{dy}{dx} = \dfrac{2t}{\cos t}, \quad \dfrac{d^2y}{dx^2} = \dfrac{2(\cos t + t\sin t)}{\cos^3 t}.$

4.8 $\dfrac{dy}{dx} = \dfrac{b\cosh t}{a\sinh t}$, undefined where $t = 0$,

$\dfrac{d^2y}{dx^2} = -\dfrac{b}{a^2\sinh^3 t}, \quad \dfrac{d^3y}{dx^3} = \dfrac{3b\cosh t}{a^3\sinh^5 t}.$

4.9 $\dfrac{dy}{dx} = \dfrac{\sin t + t\cos t}{\cos t - t\sin t}$. The curve is a spiral, so there are infinitely many places where the tangent is parallel to the y-axis.

4.10 (a) Let $f(x) = \cosh x \ (x \geq 0)$. $\dfrac{d}{dx}\left(f^{-1}(x)\right) = \dfrac{1}{\sqrt{x^2 - 1}}.$

If $g(x) = \cosh x \ (x \leq 0)$, then $\dfrac{d}{dx}\left(g^{-1}(x)\right) = -\dfrac{1}{\sqrt{x^2 - 1}}.$

(b) Let $y = \tan^{-1} x$. $\dfrac{dy}{dx} = \dfrac{1}{1 + x^2}.$

(c) Let $f(x) = e^{x^2}, \ x \geq 0$. $\dfrac{dy}{dx} = \dfrac{1}{2x\sqrt{\ln x}}.$

Let $g(x) = e^{x^2}, \ x \leq 0$. $\dfrac{dy}{dx} = -\dfrac{1}{2x\sqrt{\ln x}}.$

4.11 Let $y = \tanh^{-1} x$. $\dfrac{dy}{dx} = \dfrac{1}{1 - x^2}.$

4.12 (a) Let $f(x) = \ln x, g(x) = x$. $\dfrac{d^n}{dx^n}(f(x)g(x)) = \dfrac{(-1)^n(n-2)!}{x^{n-1}}.$

(b) Let $f(x) = e^{2x}, g(x) = x^2 - 2x + 3$. The n-th derivative of $f(x)g(x)$ is

$$2^{n-2}e^{2x}(4x^2 + (4n-8)x + (n^2 - 5n + 12)).$$

(c) Let $f(x) = e^{-x}, g(x) = x^3$. The n-th derivative of $f(x)g(x)$ is

$$= (-1)^n e^{-x}\left(x^3 - 3nx^2 + 3n(n-1)x - n(n-1)(n-2)\right).$$

Chapter 5

5.1 Let $P = (a, a^3)$. The equation of the tangent at P is $y = 3a^2x - 2a^3$. $Q = (-2a, -8a^3)$.

5.2 The distance we require is $\sqrt{3}a$.

5.3 (a) A local minimum at $x = 3$;

(b) a local maximum at $x = 0$, a local minimum at $x = 1$, a local minimum at $x = -1$;

(c) a local minimum at $x = -3$, a local maximum at $x = -1$, a local minimum at $x = 1$;

(d) a local minimum at $x = -1 - \sqrt{3}$, a local maximum at $x = -1 + \sqrt{3}$;

(e) a local minimum at $x = 0$;

(f) a local minimum at $x = 0$, a local maximum at $x = 1$, a local maximum at $x = -1$;

(g) a local minimum at $x = -1/2$;

(h) a local minimum at $x = -1$, a local minimum at $x = 1$. From the graph there is a local maximum at $x = 0$, but $f(x)$ is not differentiable there;

(i) a local maximum at $x = 0$, local minima at $x = -2, 2$ but $f(x)$ is not differentiable there;

(j) $f'(x) = \dfrac{3x^2}{(1 + x^6)^{\frac{2}{3}}} > 0$ if $x \neq 0$. $f(x)$ is strictly increasing, with a point of inflection at $x = 0$.

5.4 (a) There is a local minimum at $x = -1$, which is also the global minimum. The global maximum is at $x = 2$.

(b) There is a local minimum at $x = -1/3$, a local maximum at $x = -1$, a global minimum at $x = -2$, and a global maximum at $x = 2$.

(c) There is a local minimum at $x = -2$, which is also the global minimum. There is a local maximum at $x = 1$, which is also the global maximum.

(d) There is a local minimum at $x = -\sqrt{6}/2$, which is also the global minimum. There is a local maximum at $x = \sqrt{6}/2$, which is also the global maximum.

(e) There is a local maximum at $x = -\pi/3$, a local maximum at $x = \pi/3$, a global minimum at $x = -\pi$, and a global maximum at $x = \pi$.

(f) There are local maxima at $x = \pi/6$ and $x = 5\pi/6$, which are also global maxima. There is a local minimum at $x = \pi/2$. There is a local minimum at $x = -\pi/2$, which is also a global minimum.

(g) There is a local maximum at $x = -\pi/6$, which is also a global maximum. There are global minima at $x = \pm\pi/2$.

(h) There is a local maximum at $x = 1/\ln 3$, which is also a global maximum, and a global minimum at $x = 0$.

(i) There is a local maximum at $x = \sqrt{e}$, which is also a global maximum, and a global minimum at $x = 1$.

(j) There is a local maximum at $x = 1$, which is also a global maximum, and a global minimum at $x = -3$.

5.5 $f'(x) = -e^x \sin(e^x) = 0$ where $\sin(e^x) = 0$, i.e., $x = \ln(n\pi)$. There are 7011 solutions in the specified interval.

5.6 The two points nearest to $(0, a)$ are $\left(\pm\dfrac{\sqrt{9 + 2a^2}}{3}, \dfrac{a}{3}\right)$.

5.7 The maximum occurs at $x = \dfrac{3 - \sqrt{3}}{6}$, giving $V = \dfrac{\sqrt{3}}{9}$.

5.8 The maximum volume is attained where $r^2 = \dfrac{1}{\pi}$, giving $V_{\max} = \dfrac{8}{3\pi}$.

5.9 The minimum is where $r = \sqrt[3]{\dfrac{3000}{\pi}}$.

5.10 The cost of fuel per hour is $2048 + v^{\frac{3}{2}}$.

The minimum cost is where $v = \left(2^{12}\right)^{\frac{2}{3}} = 2^8 = 256 \, \text{km/h}$.

5.11 The equation for the Newton-Raphson iteration is

$$x_{n+1} = \frac{4x_n^5 - 8x_n^3 + 2}{5x_n^4 - 12x_n^2}.$$

We obtain the following approximations for the roots.

$$-1.9290950452, \quad -0.8478856554, \quad 2.0566715782.$$

5.12 The speed relative to the tracking station is given by

$$\frac{10^4(-9.8T_0 + 10^3)}{\sqrt{10^4 + 10^4}} \approx 634.0 \, \text{m/s},$$

where $T_0 = \dfrac{10^3 - \sqrt{10^6 - 4.(4.9).10^4}}{9.8}$.

5.13 Approximately 16,165 years.

5.14 The initial temperature is approximately $740°C$.

Chapter 6

6.1 The equation of the tangent at $(9,3)$ is $y - 3 = \dfrac{1}{6}(x - 9)$.

When $x = 9.01$, $y = 3 + \dfrac{0.01}{6} = 3.001666\ldots$.

6.2 The equation of the tangent is $y - 1 = 2\left(x - \dfrac{\pi}{4}\right)$.

When $x = \dfrac{\pi}{4} + \dfrac{\pi}{90}$ $(= 47°)$, $y = 1 + \dfrac{2\pi}{90} = 1.069813\ldots$.

6.3 Suppose that $f(x) < 0$ for all x satisfying $a < x < b$.

Suppose $a \leq x_1 < x_2 \leq b$. Then using the Mean Value Theorem gives

$$\frac{f(x_2) - f(x_1)}{x_2 - x_1} = f(c) < 0.$$

The denominator $x_2 - x_1 > 0$. Therefore $f(x_2) - f(x_1) < 0$ so the function is strictly decreasing.

6.4 The approximation is $y = e^2 + e^2(x - 2) + \dfrac{e^2}{2}(x - 2)^2$.

6.5 (a) $\sin(3x^2) = 3x^2 - \dfrac{3^3}{3!}x^6 + \dfrac{3^5}{5!}x^{10} - \cdots$;

(b) $\ln\left(2 - x^2\right) = \ln 2 - \dfrac{x^2}{2} - \dfrac{x^4}{2.2^2} - \dfrac{x^6}{3.2^3} - \cdots$;

(c) $\exp(1 + x^3) = e\left(1 + x^3 + \dfrac{x^6}{2!} + \dfrac{x^9}{3!} + \cdots\right)$;

(d) $\sqrt{x + 2} = \sqrt{2}\left(1 + \dfrac{x}{2^2} - \dfrac{x^2}{2!2^4} + \dfrac{3x^3}{3!2^6} - \dfrac{3.5x^4}{4!2^8} + \cdots\right)$;

(e) $(1 + x^2)^{-2} = 1 - 2x^2 + 3x^4 - 4x^6 + \cdots$;

(f) $\dfrac{1 + x^2}{1 - x} = 1 + x + 2x^2 + 2x^3 + 2x^4 + \cdots$;

(g) $\cos 2x = 1 - \dfrac{2^2x^2}{2!} + \dfrac{2^4x^4}{4!} - \cdots$;

(h) $\sin^2 x = \dfrac{1}{2}\left(\dfrac{2^2x^2}{2!} - \dfrac{2^4x^4}{4!} + \cdots\right)$;

(i) $\dfrac{\sin x}{x} = 1 - \dfrac{x^2}{3!} + \dfrac{x^4}{5!} - \cdots$;

(j) $\cos\left(x - \dfrac{\pi}{4}\right) = \dfrac{1}{\sqrt{2}}\left(1 + x - \dfrac{x^2}{2!} - \dfrac{x^3}{3!} + \dfrac{x^4}{4!} + \dfrac{x^5}{5!} - \cdots\right)$;

(k) $\sinh(x^3) = x^3 + \dfrac{x^9}{3!} + \dfrac{x^{15}}{5!} + \cdots$;

(l) $\cosh x - \cos x = 2\left(\dfrac{x^2}{2!} + \dfrac{x^6}{6!} + \dfrac{x^{10}}{10!} + \cdots\right)$.

6.6 (a) $e^x = e\left(1 + (x - 1) + \dfrac{(x-1)^2}{2!} + \dfrac{(x-1)^3}{3!} + \cdots\right)$;

(b) $\sin x = -(x - \pi) + \dfrac{(x-\pi)^3}{3!} - \dfrac{(x-\pi)^5}{5!} + \cdots$;

(c) $\sqrt[3]{x} = \sqrt[3]{2}\left(1 + \dfrac{1}{3.2}(x - 2) - \dfrac{1.2}{3^2.2^2.2!}(x - 2)^2\right.$

$\qquad + \dfrac{1.2.5}{3^3.2^3.3!}(x - 2)^3 - \dfrac{1.2.5.8}{3^4.2^4.4!}(x - 2)^4 + \left. \cdots\right)$;

(d) $\ln x = \ln 3 + \dfrac{(x-3)}{3} - \dfrac{(x-3)^2}{2.3^2} + \dfrac{(x-3)^3}{3.3^3} - \dfrac{(x-3)^4}{4.3^4} + \cdots$;

(e) $\dfrac{1}{x} + \dfrac{1}{x^2} = (x + 1) + 2(x + 1)^2 + 3(x + 1)^3 + \cdots$;

(f) $\dfrac{x}{1-x} = \dfrac{1}{2}\left(-1 + \dfrac{x+1}{2} + \left(\dfrac{x+1}{2}\right)^2 + \left(\dfrac{x+1}{2}\right)^3 + \cdots\right)$;

(g) $\cosh x = \cosh 2 + \sinh 2.(x - 2) + \cosh 2\dfrac{(x+2)^2}{2!}$

$\qquad + \sinh 2\dfrac{(x+2)^3}{3!} + \cosh 2\dfrac{(x+2)^4}{4!} + \cdots$;

(h) $x \sin x = -\pi(x - \pi) - (x - \pi)^2 + \pi\dfrac{(x-\pi)^3}{3!} + \dfrac{(x-\pi)^4}{3!}$

$\qquad - \pi\dfrac{(x-\pi)^5}{5!} - \dfrac{(x-\pi)^6}{5!} - \cdots$;

(i) $\ln(2 + x) = \ln 4 + \dfrac{x-2}{4} - \dfrac{1}{2}\left(\dfrac{x-2}{4}\right)^2 + \dfrac{1}{3}\left(\dfrac{x-2}{4}\right)^3 - \cdots$;

(j) $e^{x+3} = e^5\left(1 + (x - 2) + \dfrac{(x-2)^2}{2!} + \dfrac{(x-2)^3}{3!} + \cdots\right)$.

6.7 $f^{(n)}(x) = \dfrac{(-1)^{n-1}3^n(n-1)!}{(2+3x)^n}$. The Maclaurin expansion is

$$\ln 2 + \frac{3}{2}x - \left(\frac{3}{2}\right)^2 \frac{x^2}{2} + \left(\frac{3}{2}\right)^3 \frac{x^3}{3} - \left(\frac{3}{2}\right)^4 \frac{x^4}{4} + \cdots .$$

The error term is given by

$$E_n(x) = \frac{(-1)^n 3^{n+1} n!}{(n+1)!(2+3c)^{n+1}} x^{n+1} = \frac{(-1)^n 3^{n+1}}{(n+1)(2+3c)^{n+1}} x^{n+1}.$$

6.8 The Maclaurin expansion as far as the term involving x^4 is

$$\sqrt{4+x} = 2 + \frac{x}{4} - \frac{x^2}{64} + \frac{x^3}{512} - \frac{5x^4}{16384}.$$

The error term is given by

$$E_4(x) = \frac{f^{(5)}(c)}{5!} x^5 = \frac{3.5.7}{3^2.5!(4+x)^{9/2}} x^5.$$

An error bound is therefore given by

$$0 \le E_4(x) \le \frac{3.5.7}{3^2.5!.2^9}(0.1)^5 = \frac{7}{2^{17}10^5} < 0.534.10^{-9}.$$

6.9 $f^{(2n)}(x) = (-1)^n \cos x$. Now $5° = \dfrac{\pi}{36}$. An error bound is given by

$$\left| E_{2n}\left(\frac{\pi}{36}\right) \right| \le \frac{1}{(2n+1)!} \left(\frac{\pi}{36}\right)^{2n+1}.$$

We need to take $n = 2$ to be certain of six decimal places of accuracy.

$$\cos(5°) \approx 1 - \frac{\pi^2}{2.36^2} + \frac{\pi^4}{24.36^4} \approx 0.99619470$$

Chapter 7

7.1 (a) $-\dfrac{35}{2}$; (b) 2π; (c) 5; (d) 20; (e) 0; (f) 0.

7.2 (a) $x^3 + 2x^2 - 2x$; (b) $\dfrac{2(3x-1)^{3/2}}{9}$; (c) $\dfrac{x^7}{7} - x^4 + 4x$;

(d) $= \dfrac{2}{5}x^{5/2} + \dfrac{3}{4}x^4 + \dfrac{6}{11}x^{11/2} + \dfrac{1}{7}x^7$; (e) $-\dfrac{1}{x} - \dfrac{2}{x^2}$; (f) $\dfrac{3x^{2/3}}{2}$;

(g) $\sqrt{2x+3}$; (h) $\dfrac{e^{2x+3}}{2}$; (i) $-\dfrac{2^{-x}}{\ln 2}$; (j) $= \dfrac{\sinh 3x}{3}$;

(k) $-\dfrac{\cos 2x}{4}$; (l) $\tan x$; (m)$\dfrac{\sin 2x}{4} + \dfrac{x}{2}$; (n)$\dfrac{\sin 3x}{6} - \dfrac{\sin 7x}{14}$.

7.3 $F'(t) = 3t^2 e^{t^3} - 2t e^{t^2}$.

7.4 $G'(t) = -e^{\cos^2 t} \sin t - e^{\sin^2 t} \cos t$.

7.5 (a) $\displaystyle\int_1^t \frac{1}{(2x+1)^2}\,dx = \left[-\frac{1}{2(2x+1)}\right]_1^t = \frac{1}{6} - \frac{1}{2(2t+1)} \to \frac{1}{6}$ as $t \to \infty$;

(b) $\displaystyle\int_1^t \frac{1}{\sqrt{x}}\,dx = \left[2\sqrt{x}\right]_1^t = 2\sqrt{t} - 2 \to \infty$ as $t \to \infty$;

(c) $\displaystyle\int_1^t \frac{1}{(3x+2)^{\frac{2}{3}}}\,dx = \left[(3x+2)^{\frac{1}{3}}\right]_1^t = (3t+2)^{\frac{1}{3}} - 5^{\frac{1}{3}} \to \infty$ as $t \to \infty$;

(d) $\displaystyle\int_1^t e^{2-3x}\,dx = \left[-\frac{e^{2-3x}}{3}\right]_1^t = \frac{e^{-1}}{3} - \frac{e^{2-3t}}{3} \to \frac{e^{-1}}{3}$ as $t \to \infty$.

7.6 (a) For all $x \geq 1$, $\dfrac{e^{-x}}{\sqrt{x}} \leq e^{-x}$.

$\displaystyle\int_1^t e^{-x}\,dx = \left[-e^{-x}\right]_1^t = e^{-1} - e^{-t} \to e^{-1}$ as $t \to \infty$.

(b) For all x, $0 \leq e^{\sin x} \leq e$.

$\displaystyle\int_0^t \frac{e}{1+x^2}\,dx = \left[e\tan^{-1} x\right]_0^t = e\tan^{-1} t \to \frac{e\pi}{2}$ as $t \to \infty$.

(c) For all $x \geq 3$, $0 \leq \dfrac{\sqrt{x^2 - 2x - 2}}{x^3 + x + 4} \leq \dfrac{\sqrt{x^2}}{x^3} = \dfrac{1}{x^2}$.

$\displaystyle\int_3^t \frac{1}{x^2}\,dx = \left[-\frac{1}{x}\right]_3^t = \frac{1}{3} - \frac{1}{t} \to \frac{1}{3}$ as $t \to \infty$.

(d) For all $x \geq 0$, $e^{-(x^3 + x - 3)} \leq e^{3-x}$.

$\displaystyle\int_0^t e^{3-x}\,dx = \left[-e^{3-x}\right]_0^t = e^3(1 - e^{-t}) \to e^3$ as $t \to \infty$.

7.7 This is the contrapositive of Theorem 7.14. If $\int f(x)\,dx$ did converge, then by Theorem 7.14 $\int g(x)\,dx$ would also converge, contradicting the assumption.

7.8 (a) $\displaystyle\int_t^3 \frac{1}{\sqrt[3]{x-2}}\,dx = \left[\frac{3}{2}(x-2)^{\frac{2}{3}}\right]_t^3 = \frac{3}{2}\left(1 - (t-2)^{\frac{2}{3}}\right)$

$\qquad \to \dfrac{3}{2}$ as $t \to 2^+$;

(b) $\displaystyle\int_t^3 \frac{1}{x+1}\,dx = \left[\ln(x+1)\right]_t^3 = \ln 4 - \ln(t+1) \to \infty$ as $t \to -1^+$;

(c) $\displaystyle\int_t^1 \frac{x+1}{\sqrt{x}}\,dx = \left[\frac{2}{3}x^{\frac{3}{2}} + 2\sqrt{x}\right]_t^1 = \frac{8}{3} - \frac{2}{3}t^{\frac{3}{2}} - 2\sqrt{t} \to \frac{8}{3}$ as $t \to 0^+$;

(d) $\displaystyle\int_t^{\frac{\pi}{4}} \operatorname{cosec}^2 x\,dx = [\cot x]_t^{\frac{\pi}{4}} = 1 - \cot t \to \infty$ as $t \to 0^+$.

7.9 If $p = 1$, $\displaystyle\int_t^1 x^{-p}\,dx = [\ln x]_t^1 = -\ln t \to \infty$ as $t \to 0^+$. If $p \neq 1$,

$$\int_t^1 x^{-p}\,dx = \left[\frac{x^{1-p}}{1-p}\right]_t^1 = \frac{1 - t^{1-p}}{1-p} \to \begin{cases} \infty & \text{if } p > 1 \\ \frac{1}{1-p} & \text{if } p < 1 \end{cases} \quad \text{as } t \to 0^+.$$

So the improper integral diverges if $p \geq 1$ and converges if $p < 1$.

7.10 **Theorem: Comparison Test for Improper Integrals**

Suppose that $f(x)$ and $g(x)$ are continuous, and that $0 \leq g(x) \leq f(x)$, for all $a < x \leq b$. Then if the improper integral $\displaystyle\int_a^b f(x)\,dx$ converges, so does the improper integral $\displaystyle\int_a^b g(x)\,dx$, and $\displaystyle\int_a^b g(x)\,dx \leq \int_a^b f(x)\,dx$.

7.11 (a) For $0 < x \leq 1$, $0 \leq \dfrac{\sin x}{\sqrt{x}} \leq \dfrac{1}{\sqrt{x}}$.

$$\int_t^1 \frac{1}{\sqrt{x}}\,dx = [2\sqrt{x}]_t^1 = 2 - 2\sqrt{t} \to 2 \quad \text{as } t \to 0^+.$$

(b) For $-2 < x \leq 2$, $0 \leq \dfrac{e^{-x}}{\sqrt{x+2}} \leq \dfrac{e^2}{\sqrt{x+2}}$.

$$\int_t^2 \frac{e^2}{\sqrt{x+2}}\,dx = e^2 \left[2\sqrt{x+2}\right]_t^2 = e^2 \left(4 - 2\sqrt{t+2}\right) \to 4e^2 \quad \text{as } t \to -2^+.$$

Chapter 8

8.1 (a) $x\cosh x - \sinh x$; (b) $x^2 e^x - 2xe^x + 2e^x$;

(c) $\dfrac{x^2 \sinh 3x}{3} - \dfrac{2x \cosh 3x}{9} + \dfrac{2\sinh 3x}{27}$; (d) $\dfrac{x^2(\ln x)^2}{2} - \dfrac{x^2 \ln x}{2} + \dfrac{x^2}{4}$;

(e) $x^2 \sin x + 2x \cos x - 2\sin x$; (f) $\dfrac{x^2 \tan^{-1} x}{2} - \dfrac{x}{2} + \dfrac{\tan^{-1} x}{2}$;

(g) $\dfrac{x \sin 2x}{4} + \dfrac{\cos 2x}{8} + \dfrac{x^2}{4}$; (h) $\ln\left(\sqrt{x}\right)\cdot\dfrac{2x^{3/2}}{3} - \dfrac{2x^{3/2}}{9}$;

(i) $xe^x \ln x - e^x$; (j) $x\cos^{-1} x - \sqrt{1 - x^2}$.

8.2 $I = \dfrac{e^{ax}(b\sin bx + a\cos bx)}{a^2 + b^2}$.

8.3 $I = \dfrac{\sec x\tan x + \ln|\sec x + \tan x|}{2}$.

8.4 $I_n = \dfrac{e^2}{2} - \dfrac{n}{2}I_{n-1}$.

$I_4 = \dfrac{e^2}{4} - \dfrac{3}{4}$.

8.5 $I_n = \dfrac{n-1}{n}I_{n-2}$.

$I_8 = \dfrac{7}{8}I_6 = \dfrac{7\,5}{8\,6}I_4 = \dfrac{7\,5\,3}{8\,6\,4}I_2 = \dfrac{7\,5\,3\,1}{8\,6\,4\,2}I_0 = \dfrac{7\,5\,3\,1}{8\,6\,4\,2}\dfrac{\pi}{2} = \dfrac{35\pi}{256}$.

8.6 The MAPLE commands are as follows.

```
with(student); P:n->Int(x^n*exp(2*x),x=0...1);

intparts(P(n),x^n); simplify(%); value(P(0)); value(P(7));
```

8.7 $\Gamma\left(\dfrac{7}{2}\right) = \dfrac{5}{2}\Gamma\left(\dfrac{5}{2}\right) = \dfrac{5\,3}{2\,2}\Gamma\left(\dfrac{3}{2}\right) = \dfrac{15\pi}{8}$.

Chapter 9

9.1 (a) $\dfrac{(x^3 + 4)^{21}}{21}$; (b) $\dfrac{3(x^2 - 6)^{\frac{7}{3}}}{14}$; (c) $\dfrac{(x^3 + 3x - 2)^{\frac{4}{3}}}{4}$;

(d) $\dfrac{\ln(2x^3 + 5)}{6}$; (e) $\dfrac{1}{6}\tan^{-1}\left(\dfrac{x^3}{2}\right)$; (f) $\dfrac{\cos^8 x}{8}$; (g) $\dfrac{\tan^6 x}{6}$;

(h) $\dfrac{\sin(2x^3)}{6}$; (i) $-2\cos\left(\sqrt{x}\right)$; (j) $\dfrac{e^{x^2 - 2}}{2}$; (k) $\sin(\ln x)$;

(l) $= -\dfrac{e^{\cos 2x}}{2}$; (m) $2\sqrt{e^x + 3}$; (n) $\dfrac{1}{\sqrt{2}}\tan^{-1}\left(\dfrac{e^x}{\sqrt{2}}\right)$.

9.2 (a) $= \dfrac{e - 1}{3}$; (b) $\dfrac{1}{35}$; (c) $\dfrac{2}{3}$; (d) $\dfrac{195}{4}$.

9.3 (a) $\sin^{-1}\left(\dfrac{x + 2}{2}\right)$; (b) $\dfrac{x}{3\sqrt{3 - x^2}}$; (c) $\dfrac{\sin^{-1} x}{2} - \dfrac{x\sqrt{1 - x^2}}{2}$;

(d) $-\dfrac{\sqrt{1 - 9x^2}}{x}$; (e) $\sinh^{-1}\left(\dfrac{x}{2}\right)$; (f) $\sin^{-1}(x - 1) - \sqrt{2x - x^2}$;

(g) $\dfrac{x}{\sqrt{a^2 - x^2}} - \sin^{-1}\left(\dfrac{x}{a}\right)$; (h) $\dfrac{(x - 1)}{5}\sqrt{x^2 - 2x + 6} + \sinh^{-1}\left(\dfrac{x - 1}{\sqrt{5}}\right)$.

9.4 (a) $\dfrac{1}{2}\tan^{-1}\left(\dfrac{1}{2}\tan\left(\dfrac{t}{2}\right)\right)$; (b) $\ln\left|1+\tan\left(\dfrac{t}{2}\right)\right|$;

(c) $\dfrac{\pi}{3\sqrt{3}}$; (d) $\ln 3 - \ln 2$.

9.5 (a) $\cosh^{-1}\left(\dfrac{2+\sin x}{\sqrt{3}}\right)$; (b) $\sin^{-1}\left(\dfrac{e^x}{\sqrt{2}}\right)$.

Chapter 10

10.1 (a) $\ln|x-6|-\ln|x+1|$; (b) $5\ln|x-2|-2\ln|x-3|$; (c) $\tan^{-1}\left(\dfrac{x+2}{3}\right)$;

(d) $\ln|x^2+6x+10|-3\tan^{-1}(x+3)$;

(e) $x+3\ln|x^2-4x+5|+4\tan^{-1}(x-2)$;

(f) $\dfrac{x^2}{2}-4x+5\ln|x^2+6x+13|+\dfrac{21}{2}\tan^{-1}\left(\dfrac{x+3}{2}\right)$;

(g) $3\ln|x+2|+\dfrac{7}{x+2}$; (h) $5\ln|x-1|-\dfrac{10}{x-1}-\dfrac{9}{2(x-1)^2}$;

(i) $2x-12\ln|x+3|-\dfrac{7}{x+3}$;

(j) $\dfrac{1}{4}\ln|x-1|+\dfrac{3}{8}\ln|x^2+2x+5|-\dfrac{1}{4}\tan^{-1}\left(\dfrac{x+1}{2}\right)$;

(k) $x-\dfrac{5}{4}\ln|x^2+2x+5|-2\tan^{-1}\left(\dfrac{x+1}{2}\right)+\dfrac{1}{2}\ln|x-1|$;

(l) $\ln|x-1|-\dfrac{5}{x-1}-\dfrac{1}{2}\ln|x^2+2x+2|-7\tan^{-1}(x+1)$;

(m) $\dfrac{1}{2}\ln|x+1|-\dfrac{9}{4(x+1)}-\dfrac{1}{2}\ln|x-1|-\dfrac{3}{4(x-1)}$;

(n) $\dfrac{9}{2}\tan^{-1}(x-2)+\dfrac{7x-21}{2(x^2-4x+5)}$.

10.2 $-3\ln|x-1|+9\ln|x^2-4x+5|-6\tan^{-1}(x-2)-\dfrac{2x+1}{x^2-4x+5}$.

10.3 Using Taylor's Theorem gives

$$P(x)=a_0+a_1\left(x+\dfrac{b}{a}\right)+a_2\left(x+\dfrac{b}{a}\right)^2+\cdots+a_k\left(x+\dfrac{b}{a}\right)^k+E_k(x).$$

Where $E_k(x)$ is given by

$$E_k(x) = \frac{P^{(k+1)}(c)}{(k+1)!} \left(x + \frac{b}{a} \right)^{n+1}.$$

Because $P(x)$ is a polynomial of degree less than n, all its derivatives of order n and above are zero, and so the error term $E_k(x)$ is zero for $k \geq n-1$. Therefore

$$P(x) = a_0 + a_1 \left(x + \frac{b}{a} \right) + a_2 \left(x + \frac{b}{a} \right)^2 + \cdots + a_{n-1} \left(x + \frac{b}{a} \right)^{n-1},$$

so that $P(x)$ can be written as $Q \left(x + \dfrac{b}{a} \right)$, where Q is a polynomial of degree at most $n - 1$. We can now write

$$
\begin{aligned}
P(x) &= a_0 + \frac{a_1}{a}(ax + b) + \frac{a_2}{a^2}(ax + b)^2 \cdots + \frac{a_{n-1}}{a^{n-1}}(ax + b)^{n-1} \\
&= c_n + c_{n-1}(ax + b) + c_{n-2}(ax + b)^2 \cdots + c_1(ax + b)^{n-1}.
\end{aligned}
$$

Finally, dividing this equation by $(ax + b)^n$ gives

$$\frac{P(x)}{(ax + b)^n} = \frac{c_1}{(ax + b)} + \frac{c_2}{(ax + b)^2} + \frac{c_3}{(ax + b)^3} + \cdots + \frac{c_n}{(ax + b)^n}.$$

Chapter 11

11.1 $L = \displaystyle\int_1^3 \sqrt{(4t^3)^2 + (6t^5)^2}\, dt = \dfrac{733^{\frac{3}{2}} - 13^{\frac{3}{2}}}{27}.$

11.2 $L = \displaystyle\int_2^3 \sqrt{1 + \left(\dfrac{3x^2}{2} - \dfrac{1}{6x^2} \right)^2}\, dx = \dfrac{343}{36}.$

11.3 $L = \displaystyle\int_1^2 \sqrt{1 + \left(2x - \dfrac{1}{8x} \right)^2}\, dx = 3 + \dfrac{\ln 2}{8}.$

11.4 $L = \displaystyle\int_0^{\alpha} \sqrt{1 + \sinh^2 x}\, dx = \sinh \alpha = 10,$ if $\alpha = 2.9982$ (to 4 d.pl.)

11.5 $S = 2\pi \displaystyle\int_0^1 \dfrac{t^2}{2} \sqrt{t^4 + t^2}\, dt = \dfrac{\pi(2\sqrt{2} + 2)}{15}.$

11.6 $S = 2\pi \displaystyle\int_0^{\pi} \sin t \sqrt{4 \sin^2 t + \cos^2 t}\, dt = 2\pi + \dfrac{8\pi}{3\sqrt{3}}.$

11.7 $S = 2\pi \displaystyle\int_0^1 x\sqrt{1 + 0.04x^2}\, dx = \dfrac{50\pi}{3} \left(1.04^{\frac{3}{2}} - 1 \right).$

11.8 $V = \int_0^4 \frac{(4-t)^2}{6} \pi \, dt = \frac{\pi}{6} \left[-\frac{(4-t)^3}{3} \right]_0^4 = \frac{32\pi}{9}.$

11.9 Around the line $y = -1$: $V = \pi \int_{-\frac{\pi}{2}}^{\frac{\pi}{2}} \left((2 + \cos x)^2 - 2^2 \right) dx = 8\pi + \frac{\pi^2}{2}.$

Around the line $x = -\frac{\pi}{2}$: $V = 2\pi \int_0^{\frac{\pi}{2}} \left(x + \frac{\pi}{2} \right) \cos x \, dx = 2\pi(\pi - 1).$

11.10 We choose axes so that the centre of the circle is at $(0, a)$ and the tangent is the x-axis. Using the disc method gives

$$V = \pi \int_{-a}^a \left(\left(a + \sqrt{a^2 - x^2} \right)^2 - \left(a - \sqrt{a^2 - x^2} \right)^2 \right) dx = 2a^3\pi^2.$$

We now let the equation of the circle be $x^2 + (y-b)^2 = a^2$ and rotate about the x-axis. Using the disc method gives

$$V = \pi \int_{-a}^a \left(\left(b + \sqrt{a^2 - x^2} \right)^2 - \left(b - \sqrt{a^2 - x^2} \right)^2 \right) dx = 2a^2 b\pi^2.$$

11.11 $V = 2\pi \int_0^1 x(\cosh x - \sinh x) \, dx = 2\pi \int_0^1 x e^{-x} \, dx = 2\pi \left(-2e^{-1} + 1 \right).$

11.12 The total mass is $M = \int_0^1 e^x \, dx = e - 1$. The total moment about the origin is $\int_0^1 x e^x \, dx = 1$. So if the centre of mass is at distance X from the origin, $X = \frac{1}{e-1}.$

11.13 We place the figure in the first quadrant with the centre of the circle at the origin. We assume that the radius is 1, and that the density is 1. The total mass is $M = \pi/4$. The total moment about the y-axis is

$$\int_0^1 x\sqrt{1 - x^2} \, dx = \left[-\frac{1}{2}\frac{2}{3}(1 - x^2)^{\frac{3}{2}} \right]_0^1 = \frac{1}{3}.$$

So if the x coordinate of the centroid is X, we have

$$MX = \frac{1}{3} \quad \text{so} \quad X = \frac{4}{3\pi} \approx 0.424.$$

By symmetry the y-coordinate will be the same.

11.14 We place the figure in the first quadrant with the centre of the circle at the origin. Let the density be $\rho(x, y) = kx$. The total mass is given by

$$M = \int_0^1 kx\sqrt{1 - x^2} \, dx = k \left[-\frac{1}{2}\frac{2}{3}(1 - x^2)^{\frac{3}{2}} \right]_0^1 = \frac{k}{3}.$$

The total moment about the y-axis is given by

$$A = \int_0^1 kx^2\sqrt{1-x^2}\,dx = \frac{k\pi}{16}.$$

The total moment about the x-axis is given by

$$B = \int_0^1 kx\sqrt{1-x^2}.\frac{\sqrt{1-x^2}}{2}\,dx = \frac{k}{8}.$$

So if the centre of mass is at (X,Y), $X = \dfrac{3\pi}{16} \approx 0.589, Y = \dfrac{3}{8} = 0.375$.

11.15 Let the semicircle have radius a. We place it below the x-axis with its diameter between 0 and $2a$. We place the equilateral triangle, side length $2a$, above the x-axis. We assume that the density is 1. The total mass is

$$M = \sqrt{3}a^2 + \frac{\pi a^2}{2}.$$

The centre of mass will be on the axis of symmetry, so we only need its y-coordinate. The moment of the triangle about the x-axis is given by

$$T = \int_0^{\sqrt{3}a} y.2a\frac{\sqrt{3}a - y}{\sqrt{3}a}\,dy = a^3.$$

The moment of the semicircle about the x-axis is given by

$S = 2\int_0^a y\sqrt{a^2-y^2}\,dy = \dfrac{2a^3}{3}$. The total moment about the x-axis is $T - S = a^3/3$. So if the y coordinate of the centroid is Y then $MY = a^3/3$, giving $Y = \dfrac{2a}{3(2\sqrt{3}+\pi)} \approx 0.1009a$.

11.16 By symmetry the centre of mass is on the axis of the cone. The total mass of the curved surface is $M = \dfrac{5\pi}{2}\int_0^2 z^2\,dz = \dfrac{20\pi}{3}$, where we assume that the coefficient of proportionality is 1. Its moment is $\dfrac{5\pi}{2}\int_0^2 z^3\,dz = 10\pi$. If the distance of the centre of mass from the vertex is Z, $Z = \dfrac{10\pi.3}{20\pi} = \dfrac{3}{2}$. The mass of the base is $K = 2\pi\int_0^1 r\sqrt{4+r^2}\,dr = \dfrac{2\pi}{3}\left(5^{\frac{3}{2}}-8\right)$. The centre of mass is at the centre of the disc, so the total moment about the line through the vertex parallel to the base is $2K$. The moment of the whole cone about the line through the vertex is $T = 10\pi + \dfrac{4\pi}{3}\left(5^{\frac{3}{2}}-8\right)$. The total mass including the base is $W = \dfrac{20\pi}{3} + \dfrac{2\pi}{3}\left(5^{\frac{3}{2}} - 8\right)$. If the distance of the centre of mass from the vertex is D, then

$$D = \frac{10\pi + \frac{4\pi}{3}\left(5^{\frac{3}{2}} - 8\right)}{\frac{20\pi}{3} + \frac{2\pi}{3}\left(5^{\frac{3}{2}} - 8\right)} = \frac{30 + 4\left(5^{\frac{3}{2}} - 8\right)}{20 + 2\left(5^{\frac{3}{2}} - 8\right)} \approx 1.6206.$$

11.17 The mass of the cone is given by $M = \dfrac{\pi}{2}\displaystyle\int_0^2 (2 - z)^2(2 + z)\, dz = \dfrac{10\pi}{3}$.

The moment about a line through the centre of the base is

$$T = \frac{\pi}{2}\int_0^2 z(2 - z)^2(2 + z)\, dz = \frac{56\pi}{30}.$$

So if the distance of the centre of mass of the cone from the centre of the base is Z, we have

$$Z = \frac{56\pi}{30}\frac{3}{10\pi} = \frac{56}{100} = 0.56.$$

11.18 We slice the hemisphere by means of planes parallel to the base. The centroid is on the axis of symmetry of the hemisphere. We assume unit density, and unit radius. The moment about a line in the base, through the centre, is

$$T = \pi\int_0^1 z(1 - z^2)\, dz = \frac{\pi}{4}.$$

The volume of the hemisphere, and therefore its total mass, is $2\pi/3$. So if the distance of the centroid from the base is Z, $Z = \dfrac{\pi}{4}\dfrac{3}{2\pi} = \dfrac{3}{8}$.

Index